FRENCH ISLANDS

A Priced Catalogue to the Postal History of the Islands of the North and West Coasts of France

by

O. W. Newport

J. T. Whitney

O W Newport

First Edition 1981

ISBN 0 902633 73 2

Printed in Great Britain by Picton Print
Citadel Works, Bath Road, Chippenham, Wiltshire
Typeset by Chippenham Typesetting
Paper supplied by Howard Smith Papers, Bristol
Bound by Western Book Co, Maesteg

CONTENTS

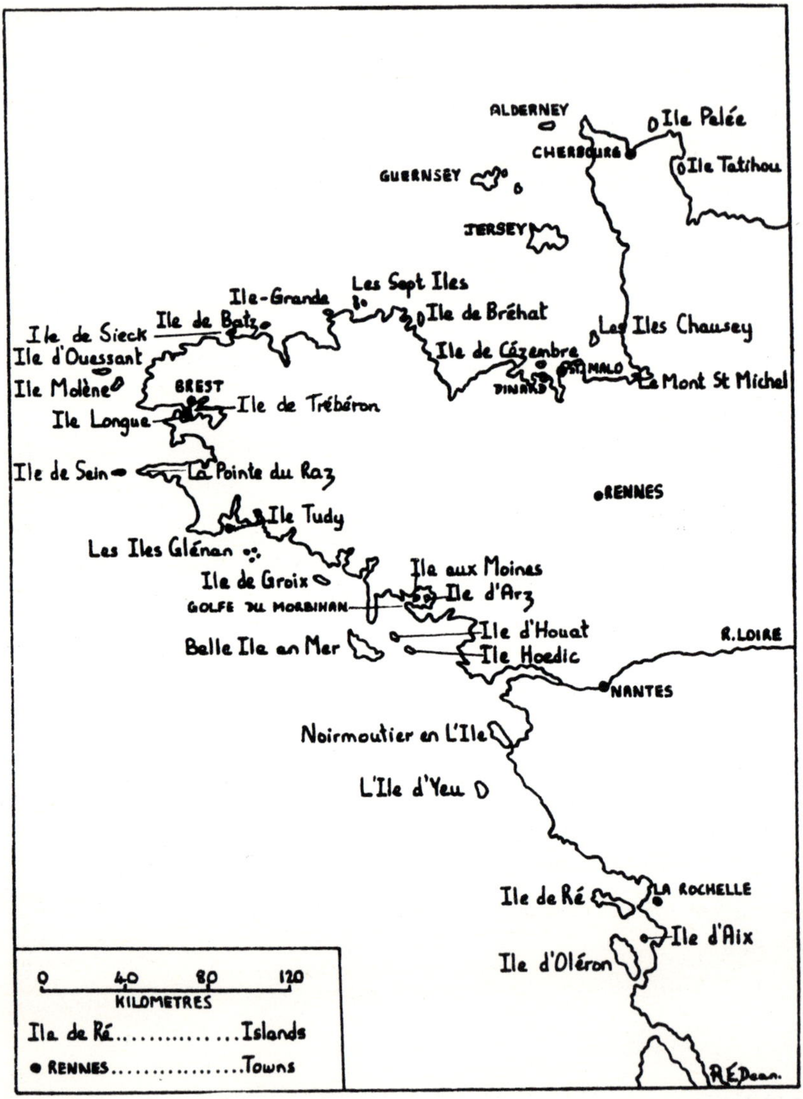
ALDERNEY
Ile Pelée
CHERBOURG
Ile Tatihou
GUERNSEY
JERSEY
Les Sept Iles
Ile-Grande
Ile de Bréhat
Ile de Sieck
Ile de Batz
Les Iles Chausey
Ile d'Ouessant
Ile de Cézembre
ST. MALO
DINARD
Le Mont St Michel
Ile Molène
BREST
Ile de Tréberon
Ile Longue
Ile de Sein
La Pointe du Raz
RENNES
Ile Tudy
Les Iles Glénan
Ile aux Moines
Ile de Groix
Ile d'Arz
GOLFE DU MORBIHAN
Ile d'Houat
R. LOIRE
Belle Ile en Mer
Ile Hoedic
NANTES
Noirmoutier en L'Ile
L'Ile d'Yeu
Ile de Ré
LA ROCHELLE
Ile d'Aix
Ile d'Oléron
0 40 80 120
KILOMETRES
Ile de Ré..............Islands
RENNES................Towns
R.E.Dean.

INTRODUCTION

To most people in Britain the term 'Channel Islands' refers to those islands which, though in close proximity to France, are under British sovereignty. There are, however, many other small islands off the north French coast which belong to the same geological formation and also several important islands off the west or Atlantic coast. Although the British Channel Islands' postal history has been intensively studied, less has been published on their French neighbours. The French islands actually offer much of interest and their stamps and postal markings reflect fully the local character of the area – tourism, ecclesiastical and secular history, piracy and privateering, fishing, agriculture and sport, wildlife, also the use of the islands to house prisoners-of-war.

The islands are listed in an east–west and then north–south direction. Markings are given following a standard pattern as follows:

1	2	3	4	5	6

1. *Reference Number.*

EH Early Handstamp.
H Handstamp.
M Machine mark, with or without slogan.
O Official mark. (It is likely that all Offices have a mark similar to Mt. St. Michel O2 or Ile Molène O1 but we list only those we have seen; the list of offices having a CORRESPONDANCE/PHILATELIQUE mark is believed complete.)
C Cachet.
FC Flight Cachet.
T Telegraphic mark.

2. *Earliest date recorded or which we have seen.*

3. *Wording.* Upper and lower case letters are reproduced as in the mark, new lines of type being indicated by a solidus (/). A blank entry in this column indicates that we have not seen an example or photograph and so cannot yet give exact wording of this mark.

4. *Description of the mark or other point of interest.* Sometimes our knowledge of the mark may be from one of the standard French reference books where it may be listed simply as a 'Gros Chiffres 2771' or 'Type A'. English readers may not be familiar with the standard French types so illustrations are given below.

5. *Colour of ink(s) known used.*

6. *Estimate of current price on card or cover.* The mark must be clear and the item clean. Prices refer to an item with low value stamps. Many cachets can be found on unused cards. They are worth approx two-thirds of price quoted. Some military cachets of a standard type (e.g. Noirmoutier C2) were used on mail from other places. Prices refer only to mail to or from the Island in question.

Price are £'s sterling, for conversion to French francs multiply by ten.

This handbook is the first attempt to produce a priced guide to this subject. As such it is based on the works listed in the Bibliography, our own collections and reports sent in by members. We are grateful to them as we are to E. H. Chambers, John Gilbert and Jill Organ for preparing illustrations. We realise that there must be many more marks from these islands and also earlier dates than those we can give at present. We hope that our readers both in Britain and in France will be able to add to the picture we have sketched. New information should be accompanied by a photocopy and will be incorporated in future editions.

O. W. Newport
6 Langton Way
London SE3 7LT

Dr. J. T. Whitney
75 Church Road
Hadleigh, Benfleet
Essex SS7 2DR

KEY TYPES

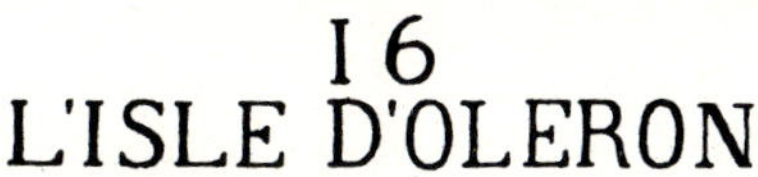

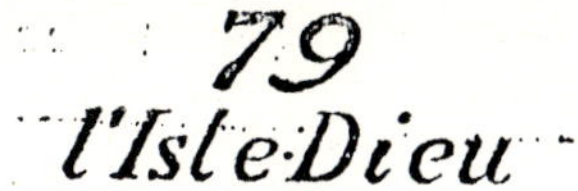

Cursive

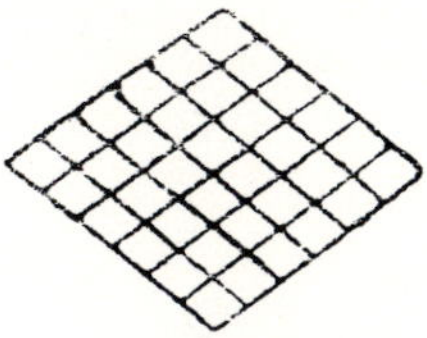
Grill

Lozenge with small numerals (Petits Chiffres)

Lozenge with large numeral (Gros Chiffres)

Type 12

Type 13

Type 15

Type 16

Type 17

Type 18

Type 24

Type 25

Type A

Type B

Early single circle

Broken single circle

Later single circle

Single circle with Department number

Early slogan

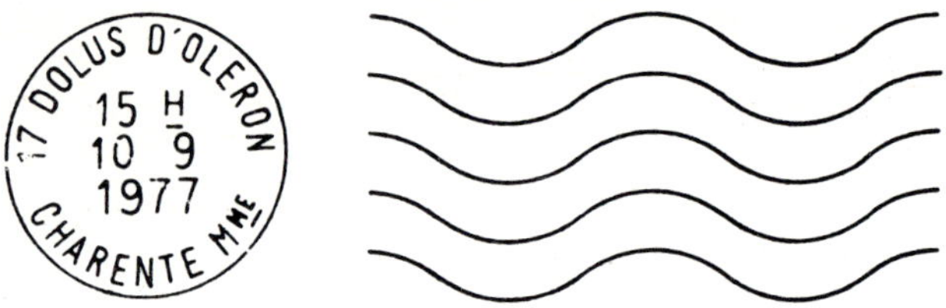

Machine with five wavy lines to right

Machine with pictorial slogan to right

Travelling

Commemorative

ILE TATIHOU (Manche)

A small island off the east coast of the Cotentin or Cherbourg Peninsula. The island is 2km from the small fishing village of St. Vaast-la-Hougue. Early postcards show views of the park and a large old house variously described as 'L'Hermitage' or 'Château d'Eau et Laboratoire'.

During the First War the island was used to house Austro-German civilian prisoners-of-war. A 1916 postcard is known with a view of Tatihou addressed to a kitchen orderly, first Division of Cavalry, Bureau Central, Paris. The sender was evidently stationed at St. Vaast-la-Hougue and serving on Tatihou. The card is endorsed at the top left 'A. Sorell, au 77 territorials 16C°, St. Vaast-la-Hougue'. The card was unstamped but also unsurcharged. There is a good ST VAAST-LA-HOUGUE/MANCHE single circle for 4th January, 1916. Part of the message reads: 'I am at this moment at St. Vaast-la-Hougue, at the Tatihou Fort, supervising the Austro-German evacuees who are all prisoners at the Fort. The duties are not hard but it is very cold on the open sea.'

Three cachets are known from this period—

C1	1915	**PLACE DE HOUGUE/ CAMP des EVACUES/Austro-Allemands**	Rectangular, boxed	Violet	20.00
C2	1916	**PREFECTURE DE LA MANCHE*/CABINET/DU/ PREFET.**	Double circle	Blue	20.00
C3	1917	**MARINE NATIONALE/ SERVICE A LA MER**	Double circle with anchor in centre	Violet	20.00

Today the island is used as an education centre for adolescents of normal intelligence who have social adaptation problems. Overprinted envelopes and notepaper inscribed 'L'Ermitage de Tatihou' are used at the Centre's offices in St. Vaast-La Hougue.

C1

C2

C3

ILE PELEE (Manche)

A small island in Cherbourg harbour which now serves as an anchor for the mole which, with the breakwater, divides the harbour from the sea. There is no post office but the island was used by the Artillerie du Front de Mer, a defence unit, during the First War.

C1 16.1.15	**Commandant de l'Ile Pelée**	Violet	20.00

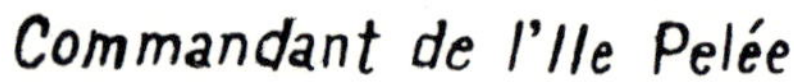

C1

LES ILES CHAUSEY (Manche)

This small archipelago of granite islands, islets and reefs is the nearest group to the British Channel Islands, being about 45km from Jersey and about 13km from Granville. According to local tradition they were once part of the Scissy Forest which was submerged by the sea in 709. Today there are 52 islands at high water and 365 at low water occupying an area of 10km from east to west and 8km from north to south. Only La Grande Ile is inhabited, though several of the larger islands have rough huts which serve as temporary shelter for fishermen; the quarries which supplied the brown granite for buildings such as Mont St. Michel are no longer worked.

The Chausey Islands figure prominently in the ecclesiastical and secular history of the area. In 1022 they were granted to the Abbey of Mont St. Michel by the Duke of Normandy. The Benedictines built an abbey which was confirmed in the possession of Mont St. Michel by Papal Bull of 1179. In 1343 the abbey was taken over by the Cordelier monks and the ordinances of the See of Coutances record the training there of several priests. The Island's fame as a seat of learning came to an end in 1543 when the abbey was abandoned after raids by Jerseymen. For two hundred years the islands were the abode of smugglers and pirates until, in 1737, the French Government built a guard house. In 1744 Jersey privateers again seized the islands, worked the quarries and shipped the stone to Jersey, Guernsey and Alderney for fortifications. The French started work on a fort in 1755 but this was destroyed by the English in July 1756. Several more forts were constructed in 1765 and a year later a strong French garrison took up quarters, thereby ending the period of rivalry over the islands' sovereignty. On 30 December 1780 Baron de Rullecourt sheltered in the archipelago for six days on his way to attack Jersey. The Fort built between 1860 and 1866 in case of English attack survives to this day. La Grande Ile, which today has a population of 109 adults, has been administered from Granville since 1804.

La Grande Ile, about 2km by 700m at its widest point is a popular resort for day trippers. An all year service is provided by the Vedettes Vertes Granvillaises and a summer service by the Vedettes Armoricaines, both from Granville. Trips are also made from St. Malo, Dinard and Jersey. The island has a substantial

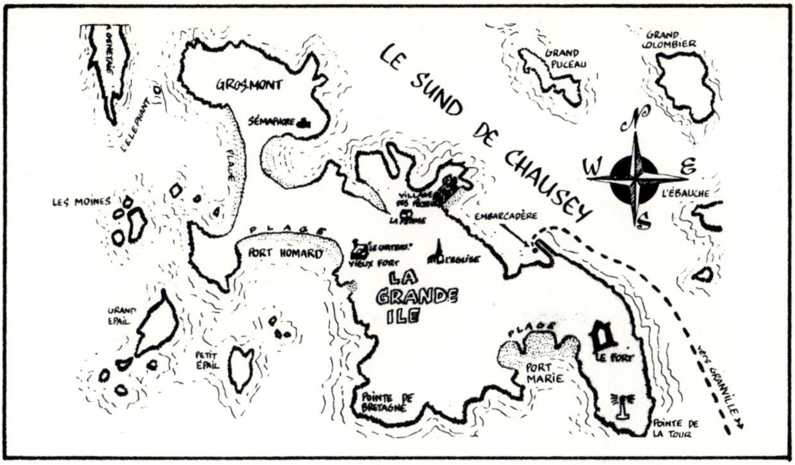

farm, a small fisherman's church, a château built in 1928 on the ruins of an earlier fort by Louis Renault, the motor magnate, a lighthouse, two shops, two hotels, a group of whitewashed cottages and curiously shaped rocks known as 'The Monks' and 'The Elephant'. Other attractions include the fishing and bird-watching; Chausey has the most important colony of Grand Cormorants in Europe.

The Cachets

There has never been a post office on the island but there is a letter box from which mail is collected twice a week and normally taken to Granville for post-marking. A number of rubber cachets have been used from the beginning of this century and, although these have no official status, they do serve to identify the mail posted there. The cachets are applied by the island priest or by the shopkeepers or hoteliers.

During the First War the island was used to house German internees. For the first months of the War mail from internees was cancelled at Granville by the office of the Commissariat of Police. Normal stamps had to be affixed, post-marked at Granville in the usual way so only the address of the sender written on the back indicates that the mail emanates from Chausey. From the end of April 1915 mail was allowed to pass without stamps and, together with mail from the La Corderie Camp on the outskirts of Granville, was marked with a different cachet. A third war cachet, known in only five examples, is believed to have been used by a defence detachment against submarines.

Mail from Chausey is known with postmarks other than those of Granville, e.g. from nearby St. Pair (1906) and further afield from St. Brieuc (1913), St. Servan-sur-Mer (1965) and Blainville-sur-Mer (1965). British international reply-paid postcards are known postmarked Hayling Island, Hampshire (1967) and Guernsey (1969). In all probability the sender did not realise that the cards were valid for use in France and gave them to visiting yachtsmen to post in the UK.

C1	1903	**RESTAURANT PIERRE MOULIN/ILES CHAUSEY**	Double circle	Blue	15.00
	1905	**As above**		Violet	15.00
C2	1904	**RESTAURANT PIERRE MOULIN/Iles Chausey**	Double circle with ship in centre	Violet	25.00
C3	1904	**HOTEL DES ILES/ CHAUSEY**	Double circle 24mm.	Violet or grey/black	8.00
	1911	**As above**		Red	8.00
C4	1907	**HOTEL des ILES/CHAUSEY**	Double circle 23mm.	Violet	8.00
	1913	**As above**		Red	8.00
C5	1907	**HOTEL/MOULIN/CHAUSEY**	Unframed, ornament below	Violet	12.00
C6	1909	**VAPEUR «ONDINE»**	Unframed, straight line	Violet	25.00
C7	1910	**PAR LE VAPEUR MONT-SAINT-MICHEL/Le . . . 19 . . .**	Unframed, two lines	Red	25.00
	1913	**As above**		Violet	25.00
C8	1910	**VAPEUR MONT-St-MICHEL ILES/CHAUSEY/Le . . . 19 . . .**	Double circle with Island wording in centre	Blue	20.00
C9	1912	**EXCURSIONS EN MER/PAR/LE VAPEUR POSTAL/ILES CHAUSEY/ GRANVILLE**	Large oval with serrated border	Black or Violet	40.00
C10	1913	**VAPEUR MONT-St-MICHEL/ILES CHAUSEY/Le . . . 19. . .**	As C7 but thinner letters & smaller star	Blue-black	20.00
C11	1913	**HOTEL Vve MOULIN/ILES/CHAUSEY/ (Manche)**	Double circle with island wording in centre	Violet	25.00
C12	1914	**ILES/CHAUSEY/ (Manche)**	Double circle (C11 cut down?)	Violet or blue	5.00
C13	1914	**COMMISSARIAT DE POLICE/ GRANVILLE (Manche)**	Double circle with design in centre, (on mail identified as from Chausey).	Violet	50.00
C14	1915	**MINISTERE DE L'INTÉRIEUR/SOUS-PREFECTURE D'AVRANCHES/DEPOT/ D'AUSTRO-ALLEMANDS/DE/LA CORDERIE/GRANVILLE/ET DE/ CHAUSEY**	Large double circle. Camp wording in centre.	Purple	50.00

C1

C3

C4

HOTEL
MOULIN
CHAUSEY

C5

VAPEUR «ONDINE»

C6

PAR LE VAPEUR MONT SAINT-MICHEL
Le .... 19 ....

C7

C8

C9

C10

C11

C12

C14

C16

C17

C18

C19

C20

C21

C22

C15	1915	**PREFECTURE DE LA MANCHE/ CABINET/DU/PREFET**	Double circle, as Tatihou C2	Blue	30.00
C16	1917	**ARTILLERIE DU FRONT DE MER-DEFENSE DES ILES CHAUSEY**	Large double circle with anchor in centre	Violet	75.00
C17	1920	**ARCHIPEL/HOTEL des ILES/CHAUSEY**	Double circle	Violet	10.00
C18	1922	**HOTEL DU FORT/ILES/CHAUSEY/ (Manche)**	Double circle with island details in centre	Blue, red or various shades of violet	8.00
C19	1927	**ARCHIPEL/des ILES/CHAUSEY**	Double circle	Black	10.00
C20	1930s	**ARCHIPEL DES ILES CHAUSEY/ (Manche)**	Double circle with serrated border	Black	10.00
	1953	**As above**		Violet	5.00
	1955	**As above**		Blue	5.00
C21	1930s	**DEBIT-TABACS/Vve MOULIN Propriétaire de l'HOTEL du FORT/ CHAUSEY/Manche**	Double circle	Green	15.00
C22	1933	**HOTEL BELLE-VUE/LEPERCHOIS, Propre/ILES CHAUSEY**	Double oval	Violet	10.00
C23	1934	**Hotel du Fort & des Iles/ Pêche Excursions/ILES CHAUSEY**	Large double oval	Red	18.00
C24	1934	**Paul MORIN/SWORS SHIP BROKER/ GRANVILLE/ILES CHAUSEY**	Circular, with anchor in centre	Black	15.00
C25	1950	**ARCHIPEL DES ILES CHAUSEY**	Large double circle, 47mm, ship in centre	Black, blue or violet	5.00
C26	1952	**ILES CHAUSEY**	Circle with serrated border, ship in centre	Green, blue or red	5.00
C27	1952	**HOTEL DU FORT ET DES ILES/ CHAUSEY/ (Manche)**	Double circle 39mm	Blue	5.00
C28	1955	**ARCHIPEL DES ILES CHAUSEY**	Double circle with Viking ship in centre	Violet, black or blue	5.00
C29	1956	**HOTEL DU FORT ET DES ILES/ CHAUSEY/MANCHE**	Double circle, 39mm, with sailing ship and island name in centre	Violet	8.00

C30	1956	**Iles Chausey-Granville par/ Vedettes Vertes Granvillaises/ "Albatros"-10 Juin 1956**	Unframed three line cachet used for a visit of the Granville Philatelic Soc. to Chausey	Violet	10.00

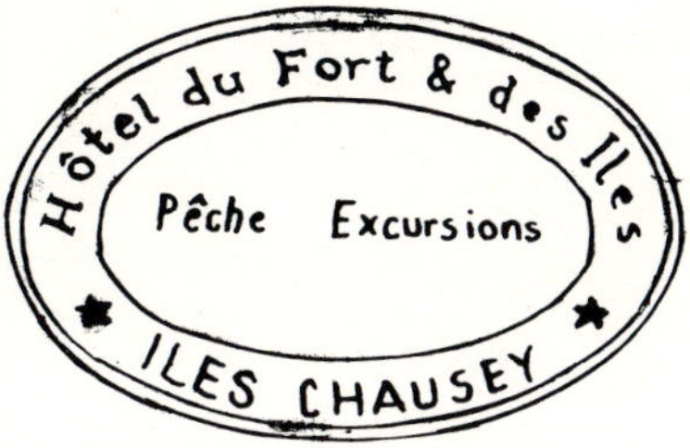

C23

C24

C25

C26

C27

C28

C29

Iles Chausey-Granville par
Vedettes Vertes Granvillaises
"Albatros" - 10 Juin 1956

C30

Service du COURRIER
Iles Chausey - Granville par
Vedettes Vertes Granvillaises
1er JOUR - " La Mouette "

C31

HOTEL BELLE-VUE
F. PERCHEY
Propr.
ILES CHAUSEY

C32

C33

VEDETTES VERTES GRANVILLAISES
SERVICE
ALBATROS
à la
MER

C34

C35

SERVICE DU COURRIER
Posté
à bord du
"Le Président QUONIAM"
VEDETTES VERTES
GRANVILLAISES

C36

HOTEL BELLE-VUE
ILES CHAUSEY

C37

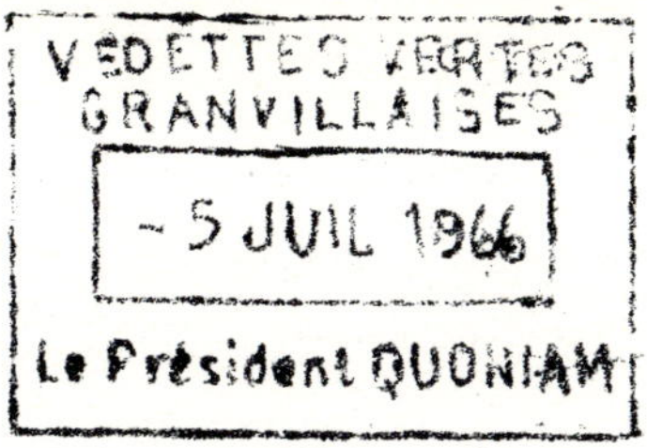

C38

C39

C31	1957	**Service du COURRIER/Iles Chausey-Granville par/Vedettes Vertes Granvillaises/1er JOUR-"La Mouette"**	Boxed cachet	Blue	10.00
C32	1957	**HOTEL BELLE-VUE/F. PERCHEY/Propr./ ILES CHAUSEY**	Double oval	Red or violet	5.00
C33	1957	**BLONDEAU/HOTEL du FORT CHAUSEY**	Large circle with ship "Richard" in centre	Green or blue	5.00
C34	1957	**VEDETTES VERTES GRANVILLAISES SERVICE/ALBATROS/à la/MER**	Double circle	Red	10.00
C35	1960	**ARCHIPEL DES ILES CHAUSEY**	As C24 but 32mm	Blue	5.00
C36	1962	**SERVICE DU COURRIER/Posté/à bord du/"Le President QUONIAM"/VEDETTES VERTES/GRANVILLAISES**	Single circle 30mm	Blue	7.00
	1964	**As above**		Violet	7.00
C37	1965	**HOTEL BELLE-VUE/ILES CHAUSEY**	Double oval (Probably C32 with centre removed when hotel changed ownership)	Blue	7.00
C38	1965	**VEDETTES VERTES/GRANVILLAISES/ date/Le Président QUONIAM**	Boxed cachet. Date in inner box	Purple or black	7.00
C39	1967	**ILES CHAUSEY (MANCHE)**	Unframed straight line cachet	Blue	10.00
C40	1967	**Plus Jamais/la Guerre**	Boxed	Blue	10.00
C41	1968	**VEDETTES VERTES GRANVILLAISES/ S.C.A.C. Successeur/TEL 136/1/3 r. Lecampion-50-GRANVILLE**	Oval	Blue	7.00
C42	1974	**ARCHIPEL DES ILES CHAUSEY**	As C35 but five pointed star at base instead of Maltese Cross	Blue	3.00
C43	1977	**HOTEL DU FORT & DES ILES/ILES CHAUSEY/par GRANVILLE-50/Téléphone: 2 Chausey/Bernard PICHARD, gérant libre.**	Boxed	Purple	1.00

C44	1977	**ARCHIPEL DES ILES NORMANDES/ILES/ CHAUSEY**	Large double circle with anchor in centre	Purple	1.00
C45	1978	**En dehors de la saison/adresser la correspondance/11, Rue des Juifs/50-GRANVILLE/Tel: (33) 50-00-03.**	Unframed. Used at Granville out of season	Black	1.00

C40

C41

C42

C43

C44

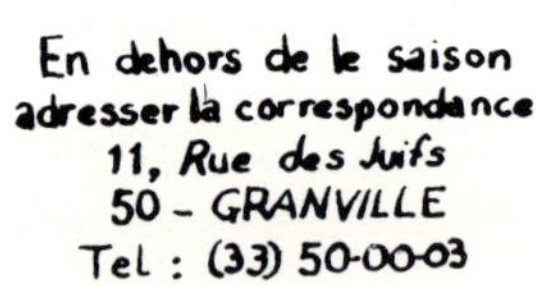

C45

The Local Stamps

There were three issues in the early 1960s when the shipping contract was in the hands of Vedettes Vertes Granvillaises. A single value appeared in 1961 followed by a set of four values a year later. Both issues were banned by the French Post Office. The second issue was held to resemble postage stamps too closely and stocks were ordered to be destroyed. The third issue, typeset labels, remained in use until 1966 when the shipping contract was taken over by Vedettes Armoricaines of Granville.

A number of examples of the 10c red are known from as late as 1964 and 1965, either as a seal or on postcards.

1961 (August 7). Typographed in sheets of 12 (3 x 4). Rouletted. Banned Sept 1961.

CH1	10c	Carmine	0.50	1.00
		Sheet of 12:	6.00	
		First Day Cover:	15.00	
		Die Proofs in blue:	20.00	
		Imperforate Proof in scarlet or emerald:		
		Single stamp	3.00	
		Sheet of 12	40.00	

1962 (January). Lithographed in Granville, perforated $11\frac{1}{2}$ line. Issued for carriage of parcels to Chausey. Banned Sept 1963.

CH2	50c	Emerald green	—	10.00
CH3	1f	Dull magenta	—	10.00
CH4	1f50	Orange	—	10.00
CH5	5f	Ultramarine	—	10.00

Vedettes Vertes Granvillaises
Tél: 1.36 GRANVILLE
0,10
Transport: Granville - Chausey

1963 (September). Typeset labels. Sheets of 8 (2 x 4). Rouletted

CH6	10c	Red	0.40	0.40
CH7	50c	Green	0.60	0.60
CH8	1f	Blue	1.00	1.00
CH9	2f	Orange	1.50	1.50
CH10	5f	Red-brown	2.50	2.50
CH11	10f	Black	5.00	5.00

Cancellations

CH1 has been seen cancelled with Cachet No.25. The first parcels issue was cancelled with a violet cancellation V.V.G./SERVICE and the second parcels issue cancelled with Cachet No.35, also in violet.

Meter Mark

A meter mark, struck in red, is now used by the Vedettes Vertes Granvillaises to publicise their trips to the Chausey islands on *Le President Quoniam*. This is applied at their offices at 1 and 3, Rue Le Campion, Granville - Value 0.50.

LE MONT ST. MICHEL (Manche)

The Mont St. Michel traces its origin to the eighth century when the Archangel Michael appeared before Aubert, Bishop of Avranches. He built a place of prayer on what had hitherto been known as Mont Tombe. A series of abbeys in the Carolingian, Romanesque and Gothic styles followed until the sixteenth century, each more splendid than the last. Granite blocks had to be brought from the Chausey Islands or Brittany and supports built up from the rocks below. During the Middle Ages the Abbey exercised a considerable influence over the surrounding area but later fell into decadence. By the late eighteenth century the Abbey had been converted into a prison. In 1874 the site passed into the care of the State. Extensive restoration and alterations were carried out. Today the Abbey is an inspiring sight, dominating the Bay over which the tide races at an astonishing speed. The Mont has a resident population of only 114 people but it is one of France's major tourist sites being visited annually by about 375,000 people. For the past century it has been connected to the Norman town of Pontorson by a causeway (la Digne), 2km in length, along which a steam tramway used to run.

A post office was first opened in 1785 remaining in use until 1797 in association with the use of the buildings as a prison. Some authorities have suggested that during this period, by analogy with other areas, the office used an unframed

MONT SAINT-MICHEL probably replaced in the 1790s by another with the office number 48/MONT SAINT-MICHEL. These marks have yet to be confirmed by actual examples. In a second period of operation the post office was open from 1 July 1849–July 1864. A small double circle and two lozenge types were used during this period, the second of which is rare. The present period of operation of the office commenced in 1880.

General Purpose Handstamps

H1	1855	**MONT St-MICHEL/48**	Type 15	Black	40.00
H2		**2052**	Small figures lozenge	Black	200.00
H3		**2418**	Large figures lozenge	Black	800.00
H4	1898	**LE MONT ST. MICHEL/MANCHE**	Type A. Usually badly struck	Black	7.50
H5	1903	**MONT ST. MICHEL/MANCHE**	Type A. Often badly struck	Black	5.00
H6	1908	**MONT ST. MICHEL/MANCHE**	Single circle	Black	2.00
H7	1934	**MONT ST. MICHEL/MANCHE**	Similar to H5 but no stop after **MONT** and more space between **MONT & MANCHE**	Black	3.00
H8	1966	**LE MONT ST MICHEL/ MANCHE**	Single circle	Black	2.00
H9	1970	**50/LE MONT ST MICHEL/MANCHE**	Single circle	Black	0.50

Machine Postmarks

M1	1937 -8	Datestamp: **MONT ST MICHEL/MANCHE**	Two single circle date stamps with five undulating lines; continuous impression	Black	3.00
M2	1947	Datestamp: **MONT ST MICHEL/MANCHE** Slogan: **LE MONT ST MICHEL/MERVEILLE/ DE L'OCCIDENT**	Two single circle date stamps with seven parallel bars, two boxed slogans; continuous impression	Black	3.50
M3	1952	Datestamp: **LE MONT-ST-MICHEL/MANCHE**	Single circle date stamp with five undulating bars to right; single impression	Black	5.00
M4	1953	As above	As M3 but bars to left of datestamp	Black	4.00
M5	1955	Datestamp: **LE MONT-ST-MICHEL/MANCHE** Slogan: **LE MONT ST. MICHEL/MERVEILLE/ DE L'OCCIDENT**	Boxed slogan; single impression	Black	2.00

H1

H2

H4

H5

H6

H7

H8

H9

M1

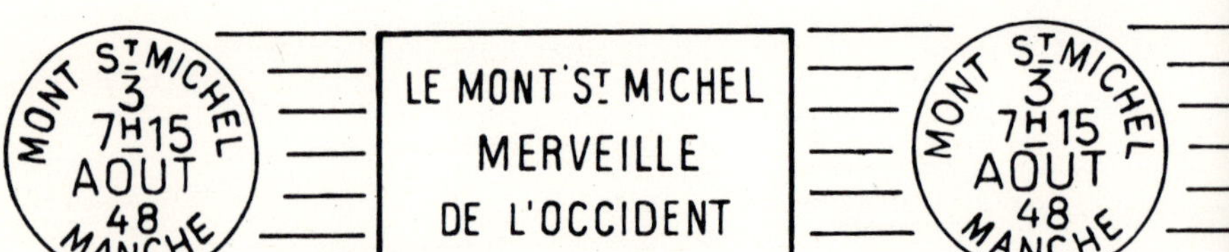

M2

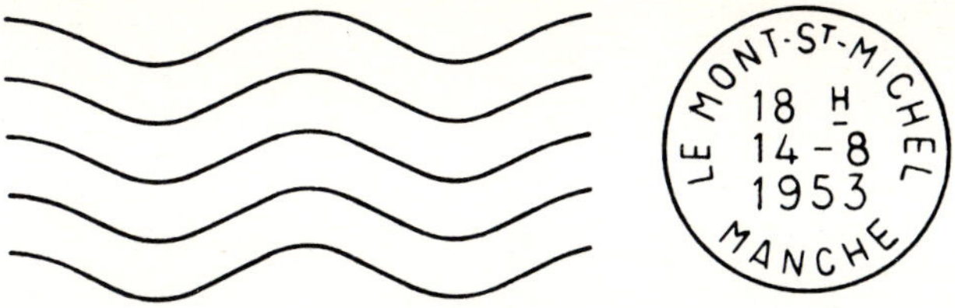

M4

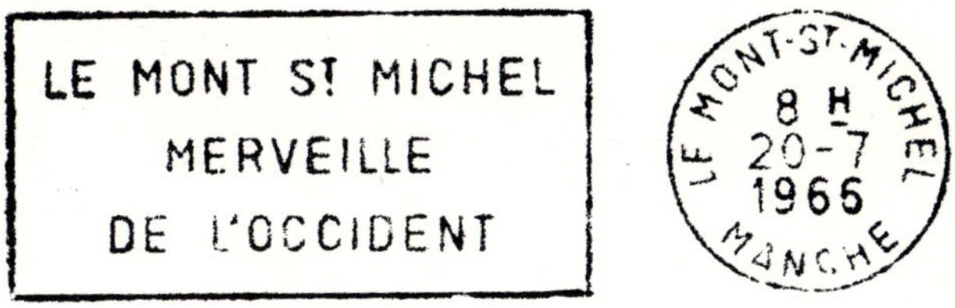

M5

M6	1967	Datestamp: **50 LE MONT ST MICHEL/MANCHE** Slogan: **LE MONT ST MICHEL/ MERVEILLE/DE L'OCCIDENT**	As M5 but datestamp with department number	Black	4.00
M7	1968	Datestamp: **50 LE MONT ST MICHEL/MANCHE** Slogan: **LE MERVEILLE/LE MONT/SAINT-MICHEL/DE L' OCCIDENT**	**LE MONT SAINT-MICHEL** simulating outline of island, date stamp to right of slogan	Black	3.00
M8	1970	**As above**	As M7 but datestamp to left of slogan	Black	0.30

Meter Marks

MM1	1930	**MONUMENTS HISTORIQUES/ ABBAYE/DU MONT SAINT MICHEL/ MONT ST MICHEL (Date)**	Unframed legend; value tablet in form of simulated stamp; machine numbers B.0208, B.0315, B.0416 and B.0895	Pink	5.00
MM2	1949	As above	Redesigned with winged letter emblem; machine number B.0232	Red	2.00
MM3	1957		Five line Camping slogan with square datestamp; machine number K2920	Red	2.00

MM4	1969	**ABBAYE/MONT ST MICHEL/1966-MILLENAIRE/date/50-MONT-SAINT-MICHEL/ (MANCHE)**	Slogan in silhouette of the Mont; machine number MG 10694	Red	3.00
MM5	1977	**HOTEL/DU/GUESCLIN/ MONT ST MICHEL/Restaurant Panoramique**	Simulated stamp; machine number P-75410	Red	0.70

Official Marks

O1	1905	**T in inverted triangle**	Used to indicate surcharge; several known to present day.	Black	3.00
O2	1966	**LE MONT/SAINT-MICHEL**	Unframed, for registration labels and official PTT mail	Black	1.00
O3	1976	**CORRESPONDANCE/PHILATELIQUE**		Black or blue	0.50

M6

MM1

O1

O2

O3

Travelling

Commemorative

Travelling Postmark

In the early part of the century a steam tramway ran across the causeway connecting Le Mont St. Michel to Pontorson. The Mont had its own railway postmark of the characteristic French type with indented sides. It was inscribed MONT ST. MICHEL à PONTORSON and is known 1902–13. Value, in black 5.00 and in red 7.50.

Commemorative Handstamp

A specially designed postmark was used on First Day covers for the stamp commemorating the thousandth anniversary of the foundation (see below). The postmark contained the legend MILLENAIRE DU MONT ST MICHEL/ PREMIER JOUR/30 AVR 66/50 LE MONT-ST-MICHEL and incorporated a head and shoulder portrait of the saint taken from the stamp design. In black, value on FDC 1.50.

Cachets

For the first three decades of this century cachets were extensively used on the Mont St. Michel, principally at the Abbey. Their purpose was to add to the novelty value of the mail rather than (as on Chausey) as a substitute postmark. Most had a short life and traces of rapid deterioration can be noted – design details became blurred and small spaces filled with ink. Only two types were dated. Cachets were evidently put on to the card at the point of sale or in advance. Unused cards are known with cachets and, in some cases where the cachet is only known as an unused card, an exact date of use cannot yet be given. Occasional examples are found with the cachet covered by the stamps confirming that cards were franked in batches and kept in stock until sold. Some cachets were in use simultaneously. The same cachet could be applied in different coloured inks and may be found on the back or front of the card. Most postmarks accompanying the cachets are the contemporary handstamps in use on the Mont (H4 or H5) though postmarks of other places are occasionally found. Publicity cachets ceased to be used in the nineteen thirties, their function being taken over by a meter mark installed at the Abbey, first recorded in 1930 (see above). Cards from 1930–1950s sometimes have a 'simulated cachet' consisting of a small double circle printed on the card in black reading 'Abbaye du Mont Saint-Michel'. Another cachet, in current use, has been noted.

The List of Cachets has been divided into five sections: General (CA), The Abbey (CB), The Ramparts (CC), The Museum (CD) and Shopkeepers (CE).

General Cachets

CA1	1913	**MONT SAINT MICHEL/L.M.**	Small double circle	Violet	5.00
CA2	1913	**SOUVENIR DU MONT SAINT MICHEL**	Double circle with picture of St. Michel	Violet	10.00
CA3	1914	**MONT-SAINT-MICHEL/Souvenir**	Double circle with serrated outer rim	Violet	5.00
CA4	1916	**SOUVENIR/DU/MONT ST MICHEL**	Small double circle with serrated rim	Violet	5.00
CA5	1918	**MONT-SAINT-MICHEL/SOUVENIR/ (Manche)**	Double circle	Violet	7.00
CA6	1921	**MONT St-MICHEL-SOUVENIR**	Small double circle	Violet	6.00
CA7	1921	**MONT St-MICHEL**	Small double circle	Violet or blue	6.00
CA8	1922	**SOUVENIR/du/MONT St-MICHEL**	Small double circle with serrated outer rim	Violet or blue	6.00
CA9	1922	**SOUVENIR DU MONT-St-MICHEL**	Small double circle with serrated outer rim	Violet	6.00
CA10	1923	**MONT ST MICHEL/RUE**	Double circle	Black	6.00
CA11	1923	**SOUVENIR du MONT-St-MICHEL**	Small double circle with serrated outer circle	Violet or green	6.00
CA12	1925	**MONT St-MICHEL**	Oval with serrated outer rim	Violet	10.00
CA13	1925	**-SOUVENIR-/du/MONT-St-MICHEL**	Small double circle, outer ring not serrated	Violet, blue or green	6.00
CA14	1927	**SOUVENIR DU MONT St-MICHEL**	Large double circle with serrated outer ring	Violet, blue or black	8.00
CA15	1929	**MONT St-MICHEL/La Merveille/ (Manche)**	Double circle	Violet	8.00
CA16	1932	**SOUVENIR DU MONT St-MICHEL**	Large double circle	Black	8.00
CA17	1933	**SOUVENIR/du/MONT-St-MICHEL/ (Manche)**	Small double circle	Violet	6.00

CA18 1930s **SOUVENIR/DU MONT/St-MICHEL**	Double circle serrated outer rim	Violet	6.00

CA1

CA2

CA3

CA4

CA5

CA6

CA7

CA8

CA9

CA10

CA11

CA12

CA13

CA14

CA15

CA16

CA17

CA18

Abbey Cachets

CB1	1902	**ABBAYE/du MONT St-MICHEL/ (Manche)**	Serrated outer circle	Violet	3.00
CB2	1905	**ABBAYE/DU/(Boxed date)/ MONT-St.MICHEL**	Rectangular with indented corners; date in inner box	Violet	4.00
CB3	1910	**ABBAYE/du MONT-St-MICHEL/ (Manche)**	Serrated outer circle	Violet or black	4.00
CB4	c1910	**ABBAYE DU MONT ST. MICHEL/ MONUMENT/HISTORIQUE**	Double circle	Violet	7.00
CB5	1911	**ABBAYE/du MONT St MICHEL/ (Manche)**	Serrated outer circle. With use top ornaments appear as squares bottom as a straight line; known used till 1920	Violet or black	2.00
CB6	1912	**ABBAYE/du/MONT/St-MICHEL/ (Manche)**	Double circle	Black or violet	5.00
CB7	1913	**ABBAYE/du MONT-St-MICHEL/ (Manche)**	Single circle very similar to CB8 but with line beneath **ABBAYE**	Black or blue	5.00

CB8	1914	**ABBAYE/du MONT-St-MICHEL/ Manche)**	As CB7 but no line beneath **ABBAYE**	Black	6.00
CB9	1915	**ABBAYE/du MONT St MICHEL/ (Manche)**	Serrated outer circle	Black	8.00
CB10	1917	**ABBAYE/du MONT St-MICHEL/ (Manche)**	Serrated outer circle	Black	8.00
CB11	1919	**ABBAYE/du MONT St-MICHEL/ (Manche)**	Serrated outer circle	Violet or Black	6.00
CB12	1921	**ABBAYE/du/MONT-St-MICHEL/ (Manche)**	Serrated outer circle	Blue, black or violet	5.00
CB13	1923	**ABBAYE DU MONT ST MICHEL/ (Manche)**	Single circle	Black	10.00
CB14	1923	**ABBAYE/du/MONT-St-MICHEL/ (Manche)**	Serrated outer circle	Blue, black or violet	5.00
CB15	1929	**ABBAYE/MONT-St-MICHEL**	Double circle	Violet	5.00
CB16	1977	**ABBAYE DU MONT-SAINT-MICHEL**	Large double circle with picture of Abbey in centre	Blue	1.00
CB17	?	**ABBAYE DU/MONT St MICHEL/ (Manche)**	Double circle with name across centre		5.00

CB1

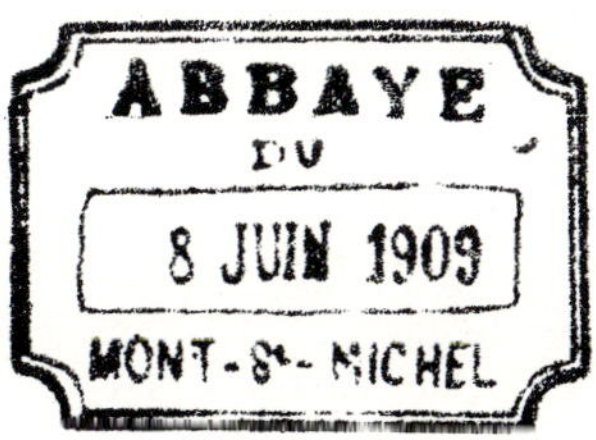

CB2

CB3

CB4

CB5

CB6

CB7

CB8

CB9

CB10

CB11

CB12

CB13

CB14

CB15

CB16

CB17

Rampart Cachets

CC1	1917	**Remparts/MONT St. MICHEL**	Double circle (CE2 with wording removed)	Violet	9.00
CC2	1918	**REMPARTS/DU MONT/SAINT-MICHEL/ (Manche)**	Serrated outer circle	Black or violet	9.00
CC3	1923	**REMPARTS/MONT-St-MICHEL/ (Manche)**	Serrated outer circle	Violet	7.00

Museum Cachets

CD1	1912	**MUSÉE DU MONT SAINT-MICHEL**	Unframed, straight line	Red	7.50
CD2	1914	**SOUVENIR/MONT/St MICHEL/ (MANCHE)/DU MUSÉE**	Double circle	Blue or red	6.00
CD3	1936?	**SOUVENIR/MONT/ST-MICHEL/ (MANCHE)/DU MUSÉE**	Double circle	Violet or black	5.00

Shop Cachets

CE1	1900s	**Magasin TURGOT/ARTICLES DE SOUVENIR/MONT-SAINT-MICHEL**	Rectangle with indented corners	Violet	8.00
CE2	1910	**Magasin des Remparts/Henri LOCHET/MONT St. MICHEL**	Double circle	Violet or black	7.00
CE3	1910	**-ETABLISSEMENTS POULARD-/MONT-St MICHEL**	Double circle	Violet or blue	8.00
CE4		**Annexe**	Used by Poulard Hotel Annex	Purple	10.00

Postmark Errors

Typical of the kind of postmark error worth watching out for are the following: H5 with missing year plug in date on 22 Aug 1909; H6 with inverted date details in centre on 25 Sept 1917; M1 with inverted date details in datestamp on 30 July 1946; CB13 with inverted details 27 Aug 1923.

Stamp Issues

It is perhaps surprising, in view of its international reputation and tourist appeal that the Mont has not featured more often on stamps issued by the French Post Office. In fact, two issues have been made. In 1929 the Mont featured in a series of tourist sites, a stamp which was re-issued in a second type two years later. In 1966 a symbolic design commemorated the millenium of the foundation. Forrester comments on the design 'St. Michel looks remarkably unconcerned when one considers that he is standing upon, and spearing, a writhing dragon. Perhaps he is thinking about his rather chic hair-do.'

CC2

CC3

CD1

CD2

CD3

CE1

CE2

CE3

CE4

Type One

Type Two

SG 473 (Yvert 260)	5 Francs, chocolate	Mint	Used
	Type One (1929) broken spire	9.50	0.12
	Type Two (1931) unbroken spire, heavier clouds	11.00	0.50
SG 1717 (Yvert 1482)	25c multicoloured	0.15	0.08

In about 1912 a group calling themselves 'Les Amis Du Mont St. Michel' issued a publicity label, perforated, printed in blue-green. The label depicted the Mont, and also the legend 'Il faut quelle Mont Saint-Michel reste une ile' (Mont St. Michel must always remain an island). The label is known affixed to the front of postcards of the period. Value 25.00.

The spiritual dominance of the Abbey of Mont St. Michel over the region in earlier times finds expression on some of the local issues of Lihou, an island off Guernsey. In the 1966 set the 4d value showed the ruins of the Lihou priory and the 6d value depicts boys with picks and shovels; both had the Mont St. Michel Abbey in the background as did the souvenir sheet which sold at 2/6. These stamps were among those overprinted for the Torrey Canyon disaster of 1967 and again for the British Postal strike of 1971.

Related Material

A 1908 advertisement letter sheet is known issued by the famous Poulard restaurant. It is printed in blue with printed imitation stamp showing St. Michael and inscribed MONT ST MICHEL/MANCHE with a 'value' of 30c. Two double circle 'postmarks' are inscribed GRANDE/TERRASSE and VEUVE/POULARD in black. Inside the letter sheet is advertising material and a view of the Mont. Value 15.00.

From 1969 comes a coloured picture postcard showing the Mont in a simulated 'stamp' which has a 'postmark' inscribed 'LES MERVEILLES DE L'OCCIDENT/LE MONT/ST MICHEL'. Value 1.00.

The tourist appeal of Mont St. Michel within the region has led to several references to the Mont appearing in postmarks of other towns:

L'Aigle	1977	**Porte de la Normandie/L'Ailge/ses/quatre jours en mai Sur la Route du Mont St. Michel**	0.50
Avranches	1924–7	**Avranches/Merveilleux/Panorama/Sur Le/Mont St. Michel** (Single circle datestamp)	4.00
	1929	As above but double circle datestamp	8.00
	1928	**Bibliotheque/D'Avranches/Exposition/Manuscripts/ Du Mt St Michel**(Single circle datestamp)	5.00
	1929	As above but double circle datestamp	8.00
	1930–8	**Avranches/Biblioth./manuscripts/Mt St Michel**	3.00
	1959–67	Pictorial Slogan as illustrated above	1.00
	1967–8	As above but datestamp **50 Avranches**	0.50
Mortain	1926–7	**Mortain/ses cascades/vue sur le/ Mont St. Michel** (slogan to left)	3.00
	1949–50	As above but slogan to right	2.00
Postorson	1970	Pictorial slogan with distant view of the Mont	0.75
St. Jean Le Thomas	1952–61	**Site Ideal/Plage/Baie du Mt/St. Michel**	2.00
	1964–6	As above	0.75
	1966	**Site Ideal/Plage de las Baie/du Mont St Michel**	0.75
	1967–8	As above but datestamp **50 St Jean le Thomas**	0.50

ILE DE CEZEMBRE (Ille-et-Villaine)

Cézembre is a rocky barren Island off St. Malo. It has no permanent residents and only three buildings situated beside a small beach. Postcard views confirm that boat trips have been made to the Island since at least the 1920s. Today pleasure trips from Dinard take 40 mins. return and allow the visitor 2½ hours on the Island. 'Le Repaire des Corsaires', ('The Corsaires' Retreat'), a small bar and restaurant, is mainly frequented by the yachting 'jet set'. During the summer, mail is carried from Dinard by a local ferry company.

Cachets

C1	1923	**ILE DE CEZEMBRE/France**	Single circle	Violet	20.00
C2	1930s	**ILE DE CEZEMBRE/France**	Single circle	Violet or brown	20.00
C3	1955	**LE REPAIRE DES CORSAIRES/ ILE DE CEZEMBRE**	Double oval with flag in centre	Blue	5.00
C4	1969	**LE REPAIRE DES CORSAIRES/ ILE CEZEMBRE**	As above but with different flag and without **DE**	Blue or violet	2.00

C1

C2

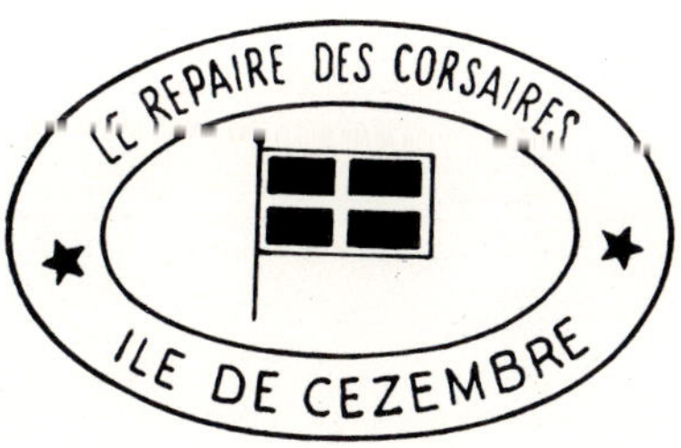

C3

C4

ILE DE BREHAT (Côtes-du-Nord)

A pair of islands, 3.5km long and 1.5km wide, which lie off the Pointe de l'Arcouest, reached by a boat trip from Paimpol. In the height of the season trips leave as regularly as every half an hour. The islands are characterised by a jagged coastline, a labyrinth of paths, pink granite rocks and low houses which have their backs to the strong winds. In the past the Bay of La Corderie was used as a port for deep-sea fishing. Today the major settlement is at La Bourg on the southern island and tourism is an important source of income. Other attractions include the St. Michael chapel, the Phare du Paon and the sensational sunsets. The population is 553. The post office was opened in 1878, according to Pothion.

Handstamps

H1	1878	**6579**	Small figures lozenge	Black	50.00
H2	1902	**ISLE-DE-BREHAT-CT.-DU-NORD**	Type 18	Black	7.00
H3	1908	**ILE DE BREHAT/COTES DU NORD**	Single circle	Black	4.00
H4	1933	**BREHAT/COTES DU NORD**	Single circle	Black	3.00
H5	1954	**BREHAT/COTES-DU-NORD**	Single circle with smaller letters and hyphens	Black	2.00
H6	1970	**22 BREHAT/COTES.DU.NORD**	Single circle	Black	0.75

Machines

M1	1924–32	Slogan: **VISITEZ/L'ILE BREHAT Datestamp: ILE DE BREHAT/COTES DU NORD**	Square framed slogan to left or right of datestamp	Black	10.00
M2	1958–60	**BREHAT/COTES DU NORD**	Five wavy bars to right	Black	2.00
M3	1960	Slogan: **BREHAT/ILE FLEURIE/A LA DOUCEUR DE VIVRE** Datestamp: **BREHAT/COTES-DU-NORD**	Framed slogan	Black	3.00
M4	1967	As M3 but datestamp; **22 BREHAT/COTES-DU-NORD**	Framed slogan to left of date stamp	Black	2.00
M5	1972	Slogan and datestamp as M4	Framed slogan to right of date-stamp	Black	2.00
M6	1974 (March 25)	Slogan: Island view with woman in native costume, **Ile Secrète/BREHAT, date** Datestamp: 22 BREHAT/COTES -DU-NORD	Framed slogan to left of datestamp	Black	0.50

Official

O1	1970	**CORRESPONDANCE/PHILATELIQUE**	Unframed	Red or black	0.50

O2	1972	**BREHAT/22-016**	Unframed. Used on Registration labels and official correspondence	Black	1.00
O3	1974	**La Receveuse**	Unframed	Red	2.00
O4	1976	**Receveuse PTT**	Unframed. Usually struck on small piece of paper stuck by Postmistress on covers. She puts her initials alongside	Red	1.00

Cachets

C1	1916	**LE SUPPLÉANT DU SOUS-INTENDANT MRE/REPUBLIQUE FRANCAISE/ ILE BRÉHAT**	Circular with allegorical figure of France in centre	Black	25.00
C2	1957	**EN SOUVENIR/DU TOUR/ DE L'ILE DE BREHAT**	Unframed three line cachet, known used with Paimpol machine. Bought on boat trip?	Blue	6.00
C3	1976	**CH. FOINANT/22870 ILE DE BREHAT/ Tel. (96) 20.00,11**	Unframed cachet applied by Tourist Agency	Black	1.00
C4	1977	**SYNDICAT D'INITIATIVE/BREHAT**	Circular 35mm across with anchor in centre	Blue	1.00
C5	1977	**SYNDICAT D'INITATIVE/BREHAT**	Circular 22mm across with anchor in centre	Black	1.00
C6	1977	**MAIRIE DE BREHAT/COTES DU NORD**	Circular with arms in centre	Black	1.00

H1

H3

H4

H5

M1

M3

M6

CORRESPONDANCE
PHILATELIQUE

O1

Receveuse PTT,

O4

C1

EN SOUVENIR
DU TOUR
DE L''ILE DE BRÉHAT

C2

CH. FOINANT
22870 ILE DE BREHAT
Tél. (96) 20.00.11

C3

C4

C5

C6

LES SEPT ILES (Côtes-du-Nord)

These seven islands are administered as a Bird Sanctuary by the Ligue Francaise pour la protection des Oiseaux. Launches from Perros-Guirec make tours round the islands. The largest island, Ile Aux Moines ('Monks' Island) has a lighthouse and a fort built by Vauban. Parties are allowed to visit this island but not the others where guillemots, auks, white and grey gulls, cormorants and gannets nest.

On 12 November 1960 the French Post Office issued a pair of stamps to promote Nature Protection. The 30c stamp multicoloured, designed and engraved in recess by Gandon, featured puffins on Les Sept Iles (SG 1504, Yvert 1274) Value: mint 0.20. used 0.12. First Day Covers were cancelled in Paris, value 1.00.

ILE GRANDE (Côtes-du-Nord)

An island off the pink Granite coast of the Corniche Bretonne. The nearest large town is Trégastel-Plage. The island, which is connected to the mainland by a road, has heaths, some rows of houses and a hotel, Hotel des Rochers. Only the marks currently in use have been recorded, but earlier ones must exist.

Handstamp

H1	1977	**22 ILE GRANDE-GA/ COTES.DU.NORD**	Single circle	Black	0.70

Official Marks

O1	1977	**T in triangle**	For surcharges	Black	1.00
O2	1977	**ILE-GRANDE-GA/ 22–932**	Unframed cachet for Registration labels and official correspondence	Black	1.00

H1

O2

ILE DE BATZ (Finistère)

Reached by a short boat excursion from Roscoff, this island is completely treeless though there is a tropical garden and there are many sandy beaches. There is a Lighthouse, a Life Saving Station and, to the north of the island, a belt of reefs. The 807 inhabitants are either sailors or farmers, the women helping to cultivate market produce and collect seaweed. The name Batz is pronounced 'Baa'. A post office was opened in 1904, according to Pothion, using the treble-circle type.

Handstamps

H1	1904	**ILE-DE-BATZ/FINISTERE**	Type B	Black	15.00
H2	1975	**29N-ILE.DE.BATZ/NORD FINISTERE**	Single circle	Black	0.70

Machines

M1	1962–4	**ILE DE BATZ/FINISTERE**	Circular datestamp with five wavy lines to right	Black	2.00
M2	1964–6	**ILE DE BATZ/NORD FINISTERE**	As above	Black	1.50
M3	1966	**29N ILE DE BATZ/NORD-FINISTERE**	As above	Black	0.75

Cachets

C1	1979	**SYNDICAT D'INITIATIVE/ Le Président/ILE DE BATZ**	Large double circle	Black	1.00

Official Marks

O1	1977	T in Triangle	For surcharges	Black	1.00
O2	1977	**ILE-DE-BATZ/29N 082**	For Registration labels & on official correspondence	Black	1.00

H1

H2

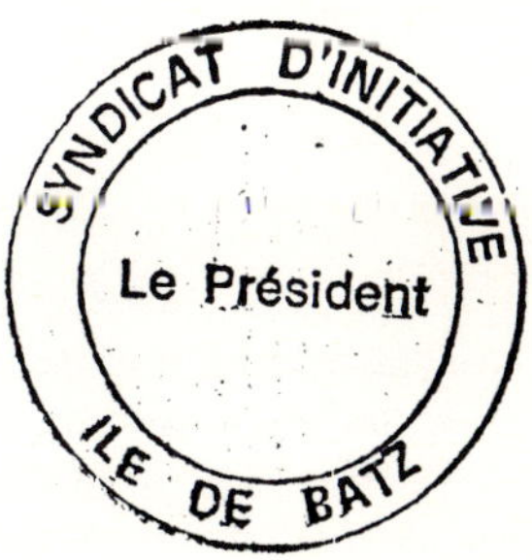

C1

O1

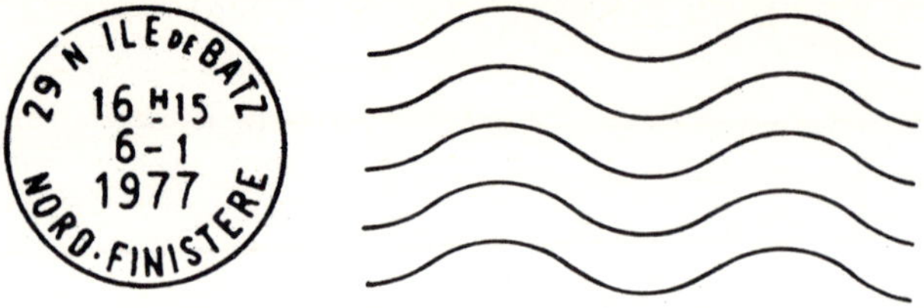

M1

ILE DE SIECK (Finistère)

The Ile de Sieck lies some five kilometres to the west of Roscoff. It has a little fishing harbour and a few houses but the population must be very small as it is not recorded in any of the standard French gazetteers. There has never been a post office on the island and no cachets used there have been recorded.

However, during the First War a camp for civilian internees was established on the island. A cover is known endorsed on the back 'Fabry Nicolos évacué, Ile de Sieck près Morlaix, (Finistere), par la Prefecture Quimper (Finistère)'. The front is marked in manuscript 'Franchise Postale Internés civils' and has a large double-circle cachet. This was probably applied at Quimper.

Cachet

C1	1915	**CONSEIL DE PREFECTURE DU FINISTÈRE/RÉPUBLIQUE/FRANÇAISE**	Double circle	Black	20.00

C1

ILE MOLENE (Finistère)

A small island with 500 inhabitants at which boats usually stop on their way from Brest to the Ile d'Ouessant. The patches of earth are so small on this archipelago that, according to local jest, a Molène cow who has her four feet in one field would graze from another and manure a third! The post office was opened on 9 August 1921.

Handstamps

H1	1921	**ILE MOLÈNE/FINISTERE**	Single circle	Black	8.00
H2	1975	**29N ILE MOLÈNE/NORD-FINISTERE**	Single circle	Black	1.00

Official Marks

01	1977	**ILE-MOLÈNE/29N-084**	Unframed cachet for Registration labels and or official correspondence	Black	1.00

H2

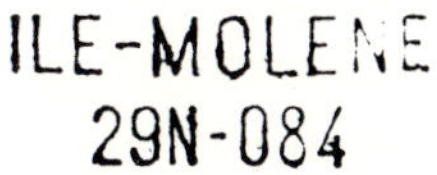

O1

ILE D'OUESSANT (Finistère)

This most westerly part of France (known to Britons as Ushant) has its own distinctive local culture and interesting traditions. By an old custom the girls propose marriage, traditional female attire consists of a short skirt and small coif, sad all night vigils are held when someone is lost at sea. The island is 7km long and 4km broad and has a population of about 1,450 people, most of whom are farmers and fishermen. Only one fiftieth of the land is able to be cultivated and, although in January and Febuary the Island has the highest mean temperature in France, the winter winds, waves and fogs have caused innumerable wrecks. In Summer months regular trips are made from Brest, Camaret or Le Conquet. There is an ornithological station attached to the Amorican Regional Museum of Natural History. The Créac'h Lighthouse marks the entrance to the Channel on the French side. Its light has a range of more than 50km and about 30,000 ships are reckoned to sight Ouessant every year. The post office was opened in 1872 as a Facteur-Boitier (office of distribution); marks used during this period are very scarce. The office began conducting full post office business in July 1912.

Handstamps

H1	1870s	**6196**	Lozenge with large figures	Black	280.00
H2	1870s	**ILE D'OUESSANT/28**	Type 24	Black	50.00
H3	1880s	**ILE D'OUESSANT/FINISTÈRE**	Type 25	Black	40.00
H4	1901	**ILE D'OUESSANT/FINISTÈRE**	Type B	Black	18.00
H5	1906	**L'ISLE D'OUESSANT/FINISTÈRE**	Type B	Black	7.50
H6	1912	**OUESSANT/FINISTÈRE**	Single circle	Black	3.00
H7	1977	**29N OUESSANT/NORD FINISTÈRE**	Single circle	Black	1.00

Machines

M1	1975	Datestamp: **29N OUESSANT/ NORD-FINISTÈRE** Slogan: ILE D'OUESSANT/ **PAYSAGES MARINE/PHARES/ MUSEE DE PLEIN AIR/ dans le Parc d'Amorique**	Framed slogan to right of datestamp	Black	0.50

Official Marks

O1	1975	**CORRESPONDANCE/PHILATELIQUE**	Unframed	Black	0.50
O2	1977	**OUESSANT/29N 155**	Unframed; for Registration labels and official correspondence	Black	1.00

Cachets

C1	1947	**MUSÉE ENEZ-EUSSA/OUESSANT**	Framed with anchors	Black	5.00
C2	1977	**MAIRIE/D'OUESSANT/(FINISTÈRE)**	Unframed	Violet	1.00

H4

H5

H6

H7

M1

O1

O2

C1

MAIRIE
D'OUESSANT
(FINISTERE)

C2

ILE LONGUE (Finistère)

The Ile Longue is a small peninsula in Brest harbour opposite the town. During the First War there was a prisoner of war camp and today there is a naval base. There is a post office and cachets are used.

Handstamps

H1	1970	**29N-BREST ILE-LONGUE-MARINE**	Single circle with anchor at base	Black	1.00

Cachets

C1	1915	**Ile-Longue**	Unframed straight line	Red	12.00
C2	1915	**PLACE FORTE DE BREST/ CENSURE MILITAIRE**	Large double circle with allegorical figure	Red	20.00
C3	1917	**87e Regiment d'Infanterie Territorial/LE COMMANDANT/ 19 Companie**	Double circle	Grey-blue	25.00
C4	1917	**REPUBLIQUE FRANÇAISE/ CENSURE/(Finistère)**	Single circle	Red	20.00
C5	1970	**MARINS-POMIERS ILE-LONGUE**	Boxed	Blue	3.00
C6	1970	**MARINE NATIONALE/MARINS-Pompiers-ILE LONGUE**	Double circle with anchor in centre	Blue	3.00
C7		**OFFICIEL/Pour le Ministre d'Etat chargé de la/Defense Nationale (Marine)/P. le Commandant de la cie/ des Marins-Pompiers/de L'ILE LONGUE/P.O. le Secretaire**	Inverted triangle	Blue	3.00

H1

Ile-Longue

C1

C2

87e Regiment d'Infanterie Territorial
LE COMMANDANT
19e Compagnie

C3

RÉPUBLIQUE FRANÇAISE
CENSURE
(Finistère)

C4

MARINS-POMPIERS ILE-LONGUE

C5

MARINE NATIONALE
MARINS-POMPIERS - ILE LONGUE

C6

OFFICIEL
Pour le Ministre d'Etat chargé de la
Défense Nationale (Marine)
P. le Commandant de la Cie
des Marine-Pompiers
de l'ILE LONGUE
P. O. le Secrétaire

C7

ILE DE TREBERON (Finistère)

Trébéron is a small island in the harbour of Brest, near the Ile Longue. Examples are known of a quarantine cachet of the First War period.

Cachet

C1	1918	**HOPITAL-LAZARET DE TRÉBÉRON/ BREST**	Double circle datestamp with anchor	Violet	40.00

LA POINTE DU RAZ (Finistère)

Although not strictly an Island, La Pointe du Raz reaches out into the ocean by a chain of reefs on the last of which is perched the Vieille Lighthouse. The spur can be toured by those firm of foot and with a head for heights. The path provides sensational views of the waves which beat unceasingly against the rocks. On a clear day the Ile de Sein can be seen in the distance. A series of cachets has been used by various restaurants and hotels at the Point.

Cachets

C1	1903	**HOTEL DE LA POINTE DU RAZ (Finistère)**	Double circle	Blue	7.00
C2	1907	**HOTEL du RAZ-de-SEIN Mme KERADENNEC/Propriétaire/ POINTE-DU-RAZ**	Double circle	Violet	4.00
C3	1908	**HOTEL DU RAZ-DE-SEIN/ LE BOUR/KERADENNEC/ Propriétaire/POINTE-DU-RAZ**	Double circle	Violet	5.00
C4	1913	**POINTE DU RAZ/LE BOUR/ KERADENNEC/HOTEL DU RAZ-DE-SEIN**	Double circle	Black	5.00
C5	1926	**HOTEL DU RAZ-DE-SEIN/ LE BOUR/KERADENNEC/ POINTE DU RAZ**	Double circle	Green or blue-green	5.00
C6	1930s	**HOTEL DE IL'ARMEN/ F. HAMON/POINTE du RAZ (Finistère)**	Double circle	Black	6.00
C7	1932	**POINTE DU RAZ/(Finistère)**	Single circle	Violet	5.00

C8	1960	**POINTE DU RAZ/(Finistère)**	Treble circle 21½mm	Red	2.00
C9	1975	**POINTE DU RAZ**	Treble circle	Blue	1.00
C10	1975	**POINTE DU RAZ/(Finistère)**	Treble circle 24mm	Blue	1.00
C11	1975	**POINTE-DU-RAZ**	Single circle 19mm	Red	1.00
C12	1975	**POINTE DU RAZ**	Single circle 21½mm	Black	1.00
C13	1975	**LA TENTATION/Madame CAM/ POINTE DU RAZ**	Double circle 30mm	Dark Blue	1.00
C14	1975	**OSTALIRI PEN AR BED/LA POINTE DU RAZ**	Double circle 36mm with lighthouse in centre	Blue	1.00
C15	1975	**Au retour/de la/Pointe du Raz**	Boxed 34 x 19mm	Dark Blue	1.00
C16	1975	**Au retour/de la/Pointe du Raz**	Boxed 34½ x 23½mm	Dark Blue	1.00
C17	1975	**Mme DONNART/Relais de l' Atlantique Hotel-Restaurant/ POINTE DU RAZ/70-06-03**	Boxed	Blue	1.00

C1

C2

C3

C4

C5

C6

C7

C8

C9

C10

C11

C12

C13

C14

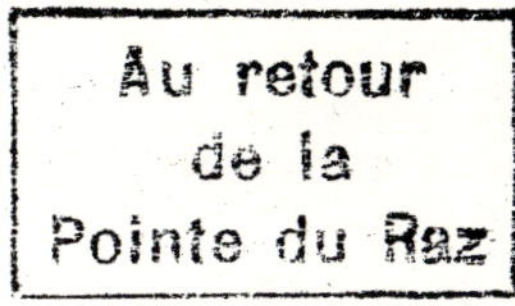

C15

C16

C17

ILE DE SEIN (Finistère)

This small bare island lies about 9km west of the Pointe du Raz and is reached by boat from Audierne. The island achieved fame during the second war because virtually all the men (130 sailors and fishermen) answered General de Gaulle's appeal and left their boats to join the Free French forces in England. Also nearly 3,000 French soldiers and sailors reached the Island and embarked for England. Twenty-seven of the Island's sailors lost their lives in the conflict and a vast War Memorial was erected after the war. General de Gaulle visited the Island in 1946 to award it the Liberation Cross. The present population is 607.

The French Ministry of Posts gives the date of opening of the post office as 1 October 1955 but as cancellations are known back to 1948 there was obviously a Facteur-Boitier some while before this.

Handstamps

H1	1948	**ILE DE SEIN/FINISTÈRE**	Broken single circle	Black	5.00
H2	1962	**ILE DE SEIN/FINISTÈRE**	Single circle	Black	2.00
H3	1967	**29S ILE DE SEIN/SUD FINISTÈRE**	Single circle	Black	1.00

Official Marks

01	1978	**ILE-DE-SEIN/ 29S-083**	Unframed two line cachet for official corres-pondence	Black	1.00

Cachet

C1	1906	**SYNDICAT/DE/L'ILE-de-SEIN**	Framed three-line	Violet	5.00

H1

H2

H3

C1

Stamp Issue

As part of the first set of Resistence Fighters Memorials a 50c stamp in indigo and light blue was issued featuring the Ile de Sein memorial. Issued on 7 April 1962, perf. 13, designed and engraved in Recess by Decaris. (S.G. 1569, Yvert 1337) Value 0.20 mint or used.

Commemorative Cancellation

A temporary post office was set up in the Town Hall using a special cancellation for the First Day of the War Memorial stamp. Value: 1.20.

ILE TUDY (Finistère)

This is actually a peninsular opposite Loctudy. It can be reached by road from Benodet or Pont l'Abbe or by boat from Loctudy. Ile Tudy itself is a pretty fishing port with 541 inhabitants.

A reading of the municipal council minutes reveals a long campaign, beginning in 1857, by the inhabitants to acquire a post office. A telegraph office was opened in 1892 and a receiving office in 1898 but full postal services were not available until the late 1920s.

Handstamps

H1	1929	**ILE-TUDY/FINISTÈRE**	Dotted single circle	Black	3.00
H2	1976	**29S-ILE-TUDY/SUD-FINISTÈRE**	Single circle	Black	0.75

Machines

M1	1962 –4	**ILE TUDY/FINISTÈRE**	Single circle datestamp with five undulating bars to right	Black	1.00
M2	1963	Slogan: **POUR VOTRE COURRIER FINISTÈRE/ PRECISEZ/NORD FINISTÈRE OU SUD FINISTÈRE**	Datestamp as above	Black	2.00
M3	1964 –6	**ILE TUDY/SUD FINISTÈRE**	Single circle datestamp with five undulating bars to right	Black	1.00
M4	1966	**29S ILE TUDY/SUD FINISTÈRE**	As above	Black	0.75

Cachet

C1	1979	**MAIRIE DE L'ILE-TUDY (Finistère)**	Double circle with allegorical figure in centre	Black	1.00

H2

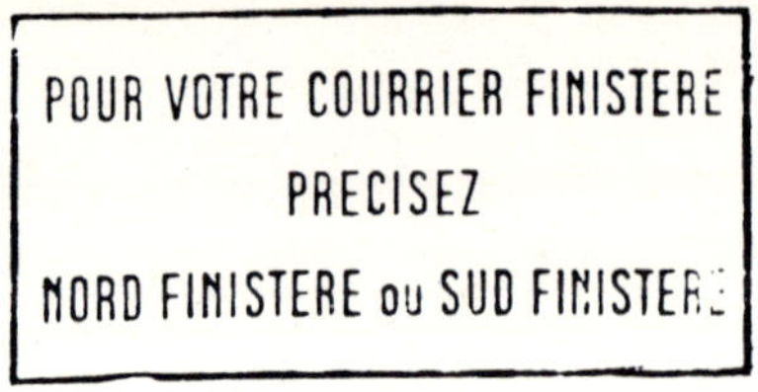

M2

LES ILES GLENAN (Finistère)

The Glénan archipelago consists of nine islands which can be reached by boat from Benodet, Quimper and Concarneau. The chief islands are Cigogne (Stork), St. Nicholas and Penfret. There are no real shops on the islands, nor is there a post office. However, on St. Nicholas there are a number of houses used during the summer months by holiday makers and a bar and store at which postcards and other souvenirs can be bought. Cachets are used on St. Nicholas and on mail from the sailing instructional school, the Centre Nautique des Glénans, on Penfret.

Cachets

C1	1975	**ILES des GLÉNAN**	Double circle with anchor in centre	Blue	2.00
C2	1975	**ST. NICHOLAS/ARCHIPEL des GLÈNAN**	Double circle with anchor in centre	Blue	2.00
C3	1975	**CENTRE INTERNATIONAL DE PLONGÉE/Secrétar-iat/114, rue Marius Anfan/92-LEVALLOIS/ LES GLÉNANS**	Single circle	Blue	2.00
C4	1975	**CENTRE NAUTIQUE/DES GLÉNANS/Quai Louis-Blériot/PARIS (16e) 520-01-40**	Unframed, four line	Violet	2.00
C5	1975	**CENTRE NATTIQUE des GLÉNANS/Quai Louis-Blériot Paris 16e**	Two line, unframed	Red	2.00
C6	1980	**Centre International de Plongée-QUIMPER/ GLÉNAN/B.P. 75-29132 CONCARNEAU**	Circular, unframed	Blue	1.00
C7	1980	**Centre International de Plongée-QUIMPER/ Director:/TREHIOU/ Jeff/CLÈNAN**	Circular, unframed	Blue	1.00
C8	1980	**Centre International de Plongée-QUIMPER/ President-Fondateur/ WEIGELE/Roger/ GLÉNAN Ecole**	Circular, unframed	Blue	1.00

C1

C2

C3

CENTRE NAUTIQUE des GLENAN
Quai Louis-Blériot PARIS-16e

C5

C6

C7

C8

ILE DE GROIX (Morbihan)

The Ile de Groix is of the same geological formation as Belle Ile though smaller, a mass of schist worn away by the sea. It is situated 14km south of Lorient from which steamers call regularly. 8km long and $2\frac{1}{2}$km wide it has a population of over three thousand. The main settlements are at Port-Tudy and Locmaria Bay.

The post office opened in November 1869 as a Facteur-Boitier and full postal facilities were available from August 1879. A succession of handstamps was used in the late nineteenth century, all of which are quite scarce. A telegraph cancellation is known on a postcard from a French soldier in 1911 and a POW cachet from 1916.

Handstamps

H1	1869	**6020**	Large figures lozenge	Black	75.00
H2	1869		Type 24	Black	35.00
H3	c1880		Type 25	Black	20.00
H4	1906	**GROIX/MORBIHAN**	Type A	Black	15.00
H5	1978	**56 GROIX/MORBIHAN**	Single circle	Black	1.00

Machines

M1	1961 -3	**GROIX/MORBIHAN**	Single circle date-stamp with five undulating bars to right	Black	1.50
M2	1963 -6	Slogan: **visites GROIX/L'ILE DES FEES/ses Plages/ Se Côte Sauvage**	Datestamp as above	Black	1.50
M3	1966	Slogan as above Datestamp: **56 CROIX Morbihan**		Black	1.00
M4	1975	**56 GROIX/MORBIHAN**	Single circle with five undulating bars to right	Black	0.75

Telegraph

T1	1911	**TELEGIE MILITRE**	Octagonal; last two letters of each word small with line beneath	Blue	16.00
	1915			Black	16.00

Cachets

C1	1915	**Regiment Territorial d'Infanterie/19 Compie/Detachment/ de/~~Belle Ile en Mer~~/ Groix**	Double circle cachet adapted to Groix by alteration from Belle Ile (Regiment indecipherable on only reported copy)	Violet	25.00
C2	1916	**VISÉ/SERVICE DE L'INTERPRÉTE/ILE de GROIX**	Oval	Violet	15.00

H4

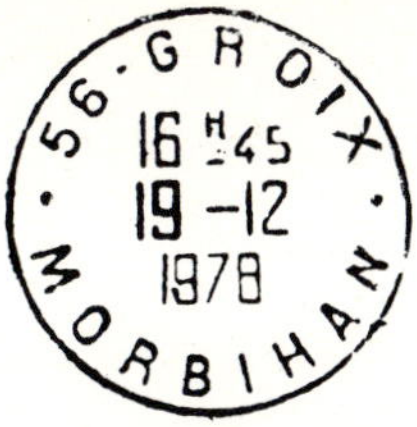

H5

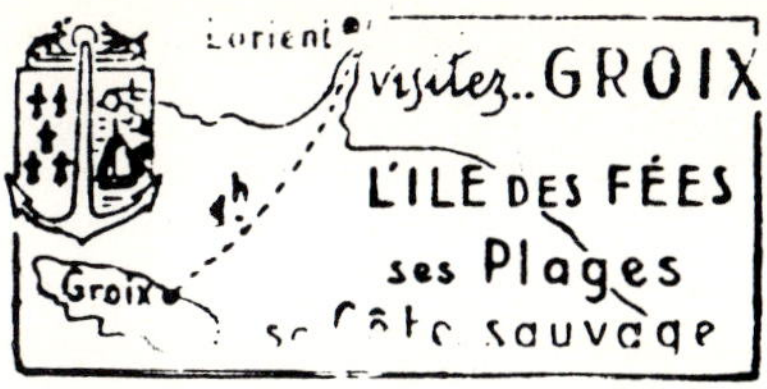

M2

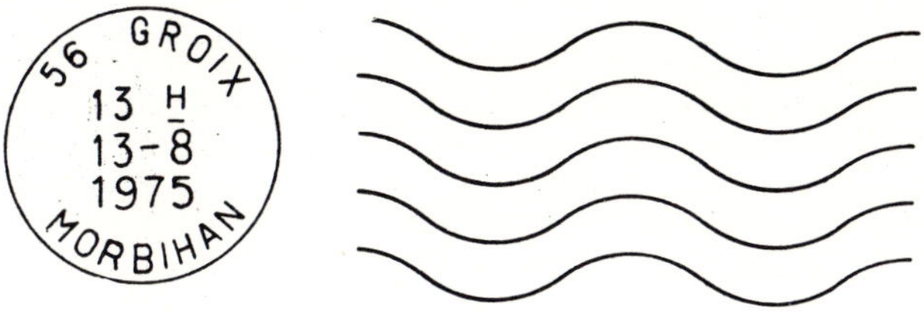

M4

T1

C1

ILE D'ARZ (Morbihan)

The Ile d'Arz (the 'z' is not pronounced) is one of the larger islands in the Gulf of Morbihan. It is about three km long and has a population of 332. The island can be reached from Vannes.

Full postal business commenced on 1 January 1974 but the office had been operated as a Facteur-Boitier since at least 1923.

Handstamps

H1	1923	**ILE D'ARZ/MORBIHAN**	Broken single circle	Black	8.00
	1960s	As above but thinner impression		Black	3.00
H2	Aug 1966	**56 ILE D'ARZ/MORBIHAN**	Broken single circle	Black	2.00
H3	1971	**56 ILE D'ARZ/MORBIHAN**	Single circle	Black	1.00

H1

H2

H3

ILE AUX MOINES (Morbihan)

The largest of the islands in the Gulf of Morbihan, this island was formerly a monastic fief. It is 6km long and is a popular holiday resort with attractive beaches and woods. The population is 588. The island is reached by boat from Vannes.

A Facteur Boitier was opened in April 1913 using a broken circle cancellation. Full postal facilities were provided from 1 July 1945, according to the Ministry of Posts. (For Ile Aux Moines on the Côtes du Nord see Les Sept Iles).

Handstamps

H1	1913	**ILE AUX MOINES/MORBIHAN**	Broken single circle	Black	12.00
H2	1959	**ILE AUX MOINES/MORBIHAN**	Single circle	Black	3.00
H3	1975	**56 ILE AUX MOINES/MORBIHAN**	Single circle	Black	1.00

Machines

M1	1966	**ILE AUX MOINES/MORBIHAN**	Single circle datestamp with five wavy bars to right	Black	3.00
M2	1966	**56 ILE AUX MOINES/ MORBIHAN**	Single circle datestamp stamp with five wavy bars to right	Black	0.75

H1

H2

H3

BELLE ILE EN MER (Morbihan)

This is the largest of the Breton islands measuring 17km x 10km at its widest point. The capital is at Le Palais on the east coast but there are also towns at Sauzon in the north and Locmaria in the south as well as 140 smaller settlements. As an outlying citadel of the French coast, the island was often attacked by the British and Dutch fleets. The British occupied it in 1761–62 until the Treaty of Paris returned it to France in 1763. Two years later many French Canadians who had refused to submit to British rule in Canada settled in the Island.

The population today is 4,328 and the Island can be reached from Lorient and Quiberon. Tourist attractions include yachting, the Côte Sauvage where the Apothicairerie Grotto is situated (so called from the way cormorants' nests used to be placed in rows in the rocky cavities like jars on the shelves of a chemist) and the small fort once owned by Sarah Bernhardt.

The post office was opened at Le Palais in 1780 and a straight line BELLE ISLE EN MER introduced. Prior to this a manuscript 'debelleisle' was used from 1725–59 and a handstruck BELLEISLE in 1778. These were probably used by a letter agent in Quiberon. Offices at Sauzon and Locmaria were opened about 1900 and in 1912 respectively. POW cachets were used in the First War and several types of private cachet are known from the Apothicairerie Grotto. The inauguration of the airmail service connecting Belle Isle with La Baule on 24 July 1938 was marked by a special cachet and signed cards.

Early Handstamps

EH1	1778	**BELLEISLE**	Single word straight line	Black	?
EH2	1781	**BELLE ISLE EN MER**	Straight line	Black	85.00
EH3	1831	**BELLE-ISLE-EN-MER/(16)**	Double circle Type 13	Black	50.00
EH4	1786	**COLONIES PAR BELLE ISLE EN MER**	Unframed oval with three fleurs des lis in centre	Black	300.00
EH5	1839	**COLONIES PAR/BELLE-ISLE SUR MER**	Two line	Black	150.00

Le Palais Handstamps

H1	1849		Grill; used with H2	Black	35.00
H2	1849	**LE PALAIS/54**	Double circle; Type 15; used with H1	Black	10.00
H3	1852	**2367**	Lozenge with small figures used with H4	Black	
H4	1852	**LE PALAIS/54**	Type 16; used with H3	Black	
H5	1863	**2771**	Lozenge with large figures; used with H6	Black	10.00
H6	1863	**LE PALAIS/(54)**	Type 17; used with H5	Black	

H7	1880s	**LE PALAIS/MORBIHAN**	Type 18	Black	8.00
H8	1990s	**LE PALAIS/MORBIHAN**	Type A	Black	5.00
H9	1911	**LE PALAIS/MORBIHAN**	Single circle	Black	2.00
H10	1978	**56 LE PALAIS/MORBIHAN**	Single circle	Black	0.60

Machines

M1	1925 -51	Slogan: **VISITEZ/BELLE ISLE/LA/SEIN NOMMÉE** Datestamp: **LE PALAIS/ MORBIHAN**	Boxed slogan to left of single circle datestamp	Black	2.00
M2	1926 -36	As above	Boxed slogan to right of single circle datestamp	Black	7.00
M3	1953		Five wavy lines	Black	1.00
M4	1958	Datestamp: **LE PALAIS MORBIHAN**	Map and ship slogan as M7	Black	1.00
M5	1965	Datestamp: **LE PALAIS/ MORBIHAN**	Bicentenial Slogan	Black	2.50
M6	1966	Datestamp: **56 LE PALAIS/MORBIHAN**	Bicentenial Slogan as M5	Black	2.50
M7	1966	Datestamp: **56 LE PALAIS/MORBIHAN**	Map and ship slogan	Black	0.50

Sauzon Handstamps

H1	1903	**SAUZON/MORBIHAN**	Type B	Black	7.50
H2	1978	**56 SAUZON/MORBIHAN**	Single circle	Black	1.00

Locmaria Handstamp

H1	1978	**56 LOCMARIA/MORBIHAN**	Single circle	Black	1.00

Cachets

C1	1978	**MAIRIE/DE LOCMARIA-BELLE-ILE 56**	Unframed, two line	Black	1.00
C2	1978	**MAIRIE DE LOCMARIA-BELLE-ILE (Morbihan)**	Double circle with allegorical figure of France	Black	1.00

Belle Isle Cachets

C1		**GROTTE DE L'APOTHICARERIE/ HOTEL/BELLE-ILE-EN-MER**	Circular		5.00
C2	1904	**GROTTE DE L'APOTHICARERIE/ BELLE-ILE/EN/MER**	Circular with serrated rim	Violet	5.00
C3	1905	**GROTTE DE L' APOTHICARERIE/Belle-ile-en-Mer**	Single circle	Violet	3.50
C4	1915	**PRISONNIER DE GUERRE du/Depot des Prison-nieres de Guerre/ BELLE-ILE**	Straight line	Black	8.00

C5	1915	**VV/par l'interprète du/Depot des Prisonnieres de Guerre/ BELLE-ILE**	Boxed	Violet	15.00
C6	1915	**DEPOT des PRISONNIERES de GUERRE/Le Commandant**	Double circle with allegorical figure in centre	Violet	12.00
C7	1919	**MARINE FRANCAISE/ SERVICE DE LA MER**	Double circle with anchor in centre	Blue	10.00
C8	1976	**A LA GROTTE/L' APOTHICARERIE/BELLE-ILE/ EN MER**	Double oval	Violet	2.00

Flight cachets

FC1	24 July 1938	**1er Vol/postal/Belle-Ile/en Mer-/La Baule/ 24 Juillet 1938**	First flight cachet with aeroplane	Violet or Black	16.00
FC2	24 July 1938	**1er Vol/postal/La Baule-/ Belle-Ile/en Mer/24 Juillet 1938**	First flight cachet with aeroplane	Red or Violet	16.00

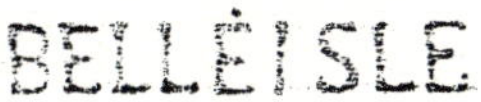

EH 1

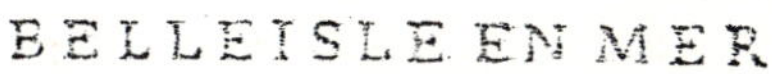

EH 2

EH 3

EH4

Le Palais H5

Le Palais H6

Le Palais H9

Le Palais H11

Le Palais M1

Le Palais M5

Le Palais M7

Sauzon H1

Sauzon H2

Locmaria H1

C1

C2

C3

PRISONNIER DE GUERRE

C4

C5

C6

C7

C8

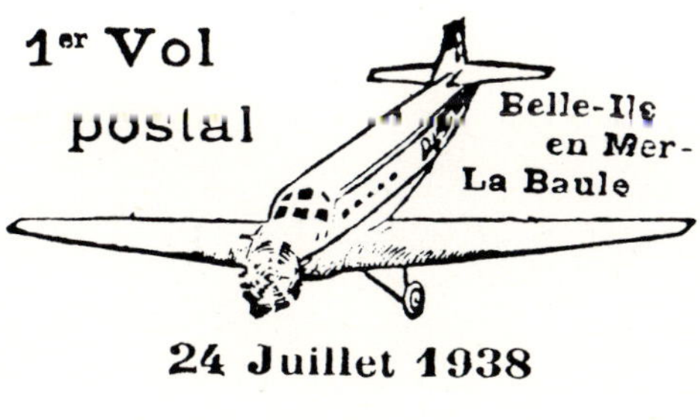

FC1

ILE D'HOUAT (Morbihan)

Houat lies 12km south-east of Quiberon and 14km north-east of Belle-Ile. It is an island village of about 450 people and has some fine beaches. A post office was opened by 1975. A postcard of 1911 is known with a double circle cancellation but the usage was telegraphic and the stamp was cancelled at Quiberon.

Handstamps

H1	1975	**56 HOUAT/MORBIHAN**	Single circle	Black	1.00

Telegraph

T1	1911	**ILE-D'HOUAT/MORBIHAN**	Type A	Black	10.00

H1

T1

ILE HOEDIC (Morbihan)

Hoedic is a small island 7km south-east of Houat. On the coast are several interesting caves and two menhirs, in one of which is a niche containing a statue of the Virgin Mary. The population is about 150.

A post office was opened in 1975 but the cancellation is not normally used to cancel the stamps. Mail is taken to Quiberon for cancellation.

Handstamps

H1	1975	**56 HOEDIC/MORBIHAN**	Single circle	Black	1.00

NOIRMOUTIER EN L'ILE (Vendée)

This island, now joined to the mainland by a road bridge, lies opposite Beauvoir-sur-Mer. The island's attractions include a castle, an ancient church and several fine beaches. The population today is just over four thousand.

The Conseil des Postes decided to open a post office on 1 January 1787, but the first mark is known from 1785, probably used by the distributor at Beauvoir. During the Revolutionary period the Armée de l'Ouest changed the island's name to L'Isle de la Montagne and a handstamp using this designation was in use for a short while. By 1798 the earlier name had been brought back into use.

Handstamps

H1	1785	**NOIRMOUTIER**	Straight line	Black	50.00
H2	1792	**79/NOIRMOUTIER**	Two line	Black	40.00
H3	1794	**79/L'ISLE DE LA MONTAGNE**	Two line	Black	150.00
H4	1798	**79/MOIRMOUTIER**	Two line	Black	30.00
H5	1799	**P.79.P./NOIRMOUTIER**	Two line	Black	30.00
H6	1806	**DEB.NOIRMOUTIER**	Straight line	Black	50.00
H7	1820	**79/NOIRMOUTIER**	Two line	Black	25.00
H8	1834	**NOIRMOUTIER/(79)**	Type 12	Black	25.00
H9	1846	**NOIRMOUTIER/(79)**	Type 13	Black	50.00
H10	1846	**NOIRMOUTIERS/(79)**	Type 15 used with H11	Black	40.00
H11	1849		Grill used with H10	Black	
H12	1852	**2285**	Lozenge with small figures; used with H10	Black	25.00
H13	1863	**2673**	Lozenge with large figures; used with H10	Black	15.00
H14	1863	**NOIRMOUTIER/(79)**	Type 15 used with H13	Black	20.00
H15	1869	**NOIRMOUTIERS/(79)**	Type 16 used with H13	Black	15.00
H16	1874	**NOIRMOUTIERS/(79)**	Type 17 used with H13	Black	15.00
H17	1876	**NOIRMOUTIER/79**	Type 18	Black	7.50
H18	1900	**NOIRMOUTIER/VENDEE**	Type A	Black	5.00
H19	1907	**NOIRMOUTIER/VENDEE**	Single circle 26mm	Black	2.00
H20	1932	**NOIRMOUTIER/VENDEE**	Single circle 28mm	Black	3.00
H21	1949	**NOIRMOUTIER/VENDEE**	Single circle 27mm	Black	2.00
H22	1967	**NOIRMOUTIER-EN-L'ILE VENDEE**	Single circle	Red	4.00
H23	1978	**85 NOIRMOUTIER-EN-L'ILE/VENDEE**	Single circle	Black	0.75

Machines

M1	1934	Datestamp: **NOIRMOUTIER VENDEE.** Slogan: **VENEZ A/ NOIRMOUTIER/L'ILE/MOMOSAS**	Boxed slogan to right of datestamp	Black	8.00
	1935 -9	As above	Slogan to left of datestamp		6.00
M2	1934	Datestamp: **NOIRMOUTIER/ VENDEE.** Slogan: **VENEZ A NOIR-/MOUTIER/EN BATEAU/ OU EN AUTO**	Boxed slogan to right of datestamp	Black	7.00
M3	1951 -3	**NOIRMOUTIER/VENDEE**	Five wavy bars	Black	2.50
M4	1954 -7	Slogan: **ILE AUX MINOSAS/ PLAGES ET BOIS/PASSAGE DU GOIS.** Datestamp: **NOIRMOUTIER/VENDEE**		Black	1.50
M5	1957 -64	Slogan as above. Datestamp: **NOIRMOUTIER EN L'ILE/ VENDEE**		Black	1.00
M6	1964 -66	Slogan: **DETENTE NOIRMOUTIER/ Ile Oceane.** Datestamp: **as M5**	Pictorial slogan to right of datestamp	Black	1.00
M7	1966	Slogan as above, datestamp; **85 NOIRMOUTIER EN L'ILE/ VENDEE**	Pictorial slogan to right of datestamp	Black	1.00
M8	1976	Datestamp: **85 NOIRMOUTIER/ EN L'ILE/VENDEE.** Slogan: **NOIRMOUTIER ILE OCEANE/ NOIRMOUTIER**	Dated pictorial slogan to left of datestamp	Black	0.75

La Gueriniere : Handstamp

H1	1978	**85 LA GUERINIERE/VENDEE**	Single circle	Black	1.00

: Machines

M1	1979	**85 LA GUERINIERE/VENDEE**	Single circle with five wavy bars to right	Black	1.00

Noirmoutier Cachets

C1	1915	**Depôt d'Étrangers/de/NOIRMOUTIER (Vendée)**	Boxed	Violet	20.00
C2	1967	**PREFECTURE DE LA VENDÉE CONTROLE**	Double circle with V; also used on Internee mail from Ile d'Yeu	Violet	15.00
C3	1967	**HOPITAL HOSPICE/ NOIRMOUTIER (Vendée)**	Double circle	Blue	3.00
C4	1967	**HOPITAL-HOPICE- NOIRMOUTIER**	Straight line	Blue	3.00
C5	1978	**NOIRMOUTIER-EN-L'ILE/HOPITAL/ LOCAL/DE/VENDEE**	Double circle	Blue	2.00
C6	1978	**NOIRMOUTIER EN L'ILE/HOPITAL/ RURAL/DE/VENDEE**	Double circle	Blue-black	2.00
C7	1978	**HOPITAL LOCAL DE NOIRMOUTIER**	Straight line	Blue	2.00
C8	1978	**OFFICE DE TOURISME/B.P.63 Tel.: 39.10.71/85330 NOIRMOUTIER**	Double circle	Black	0.85

H1

H2

H3

H4

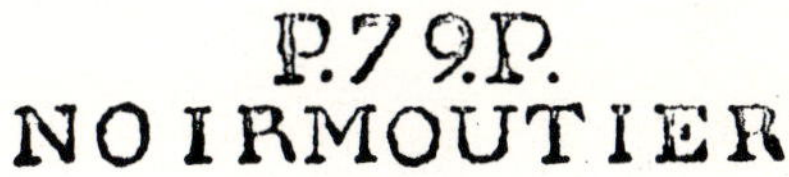

H5

H6

H7

H14

H18

H19

H20

H21

H22

H23

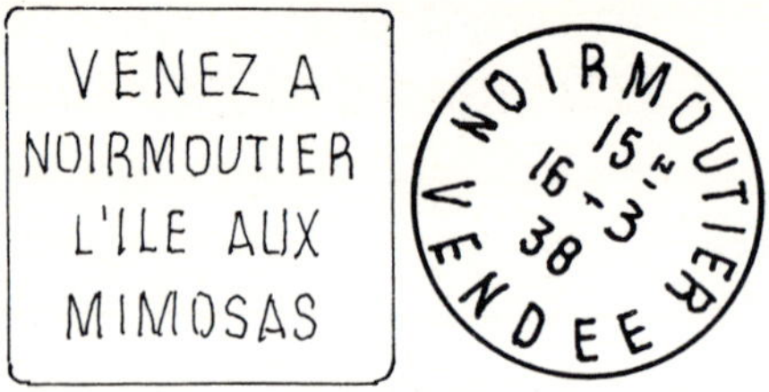

M1

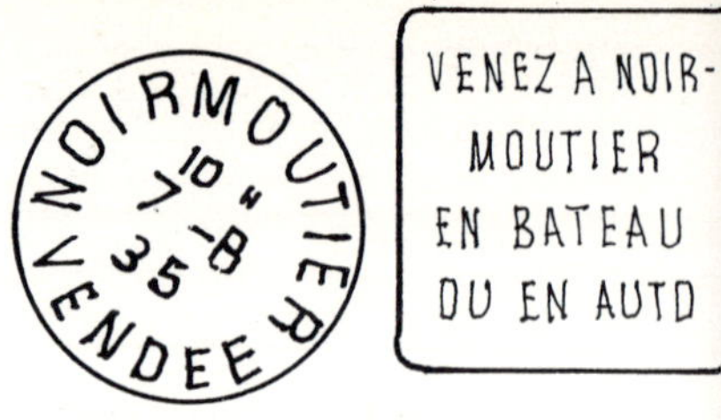

M2

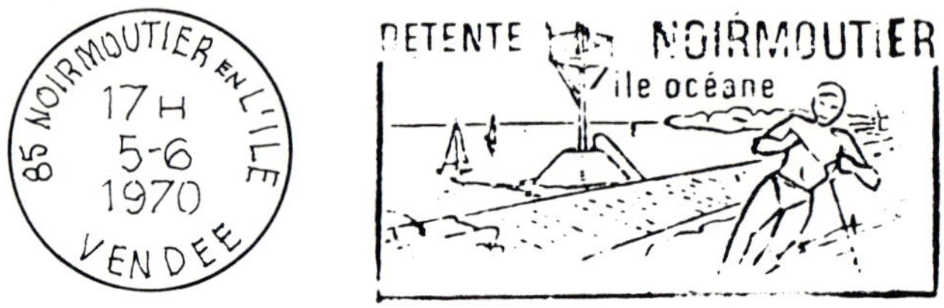

M7

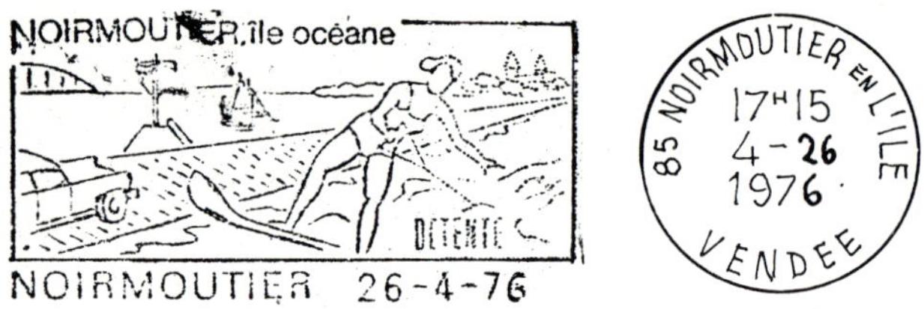

M8

C2

C3

HOPITAL-HOSPICE-NOIRMOUTIER

C4

C5

C6

HOPITAL LOCAL DE NOIRMOUTIER

C7

Dépôt d'Étrangers
de
NOIRMOUTIER (Vendée)

C1

L'ILE D'YEU (Vendée)

This island was known as L'Isle Dieu until December 1875. It lies off the port of Fromentine from which a car ferry connects with the island's port, Joinville. The island's attractions include the fishing, the coastline, an old castle and a lighthouse. The present population is 4,766.

A post office was opened on 15 July 1803, 26 Messidor Year II in the Revolutionary calendar. The first recorded mark is a straight line followed by two other undated markings. In this period, the island was controlled from St. Gilles Sur Vie but on 26 September 1833 control passed to Beauvoir-Sur-Mer after a request to this effect had been sent to the Postal Administration. First War internee mail is known cancelled with the same marking as used on Noirmoutier. A special cancellation was used in 1961 for the launching of the island's new ferry, the President August Durand.

Handstamps

H1	1819	**ILE DIEU**	Straight line	Black	80.00
H2	1819	**79/L'Isle.Dieu/St. GILLES.S.VIE**	Three line	Black	70.00
H3	1830	**79/L'Isle.Dieu**	Two line	Black	35.00
H4	1841	**L'ISLE-DIEU/(79)**	Type 15	Black	17.50
H5	1849		Grill; used with H4	Black	20.00
H6	1852	**1547**	Lozenge with small figures, used with H4	Black	40.00
H7	1863	**1835**	Lozenge with large figures, used with H4, H8 or H9	Black	30.00
H8	1869	**L'ISLE-DIEU/(79)**	Type 16, used with H7	Black	15.00
H9	1876	**L'ILE-D'YEU/(79)**	Type 17, used with H7	Black	12.00
H10	1905	**L'ILE D'YEU/VENDEE**	Type A	Black	7.50
H11	1912	**L'ILE D'YEU/VENDEE**	Single circle 27mm	Black	3.00
H12	1936	**L'ILE D'YEU/VENDEE**	Single circle 26½mm	Black	3.00
H13	1948	**L'ILE D'YEU/VENDEE**	Single circle larger letters 27½mm	Black	2.00
H14	1966	**L'ILE D'YEU/VENDEE**	Single circle 28mm	Black	1.00

Machine Cancellations

M1	1939	Slogan: **ILE D'YEU SES/BEAUX SITES/SON PORT/SES PLAGES**	Slogan to right of datestamp worded 'Ile d'Yeu'	Black	7.50
M2	1954	Slogan: **L'ILE D'ILE D'YEU/REPOS/ LUMIERE-BEAUTE**	Datestamp 'L'ile d'Yeu'	Black	1.50
M3	1966	Slogan as above Datestamp: **85 L'ILE D'YEU/ VENDEE**	Slogan to left of datestamp	Black	0.75

Official Marks

O1	1978	**CORRESPONDANCE/PHILATELIQUE**	Unframed	Black	1.00

Commemorative Handstamp

CH1	12 June 1961	**PRESIDENT AUGUSTE DURAND/ ILE D'YEU**	Pictorial cancellation	Black	4.00

Cachets

C1	1915 –17	**PREFECTURE DE LA-VENDEE/ CONTROLE**	As C1 of Noirmoutier	Red or violet	15.00
C2	1979	**SYNDICATE/DE L'ILE D'YEU/ D'INITIATIVE**	Double circle	Blue	1.00

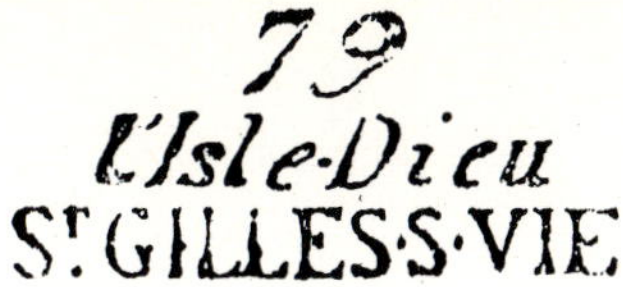

H2

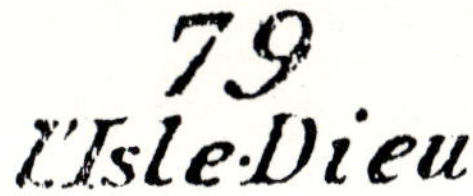

H3

H4

H7

H8

H10

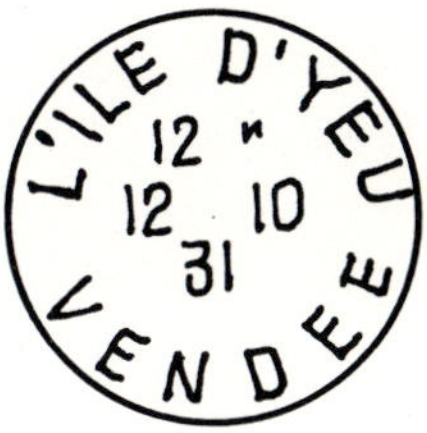

H11

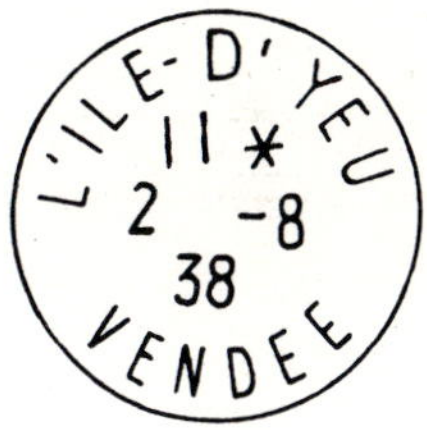

H12

H13

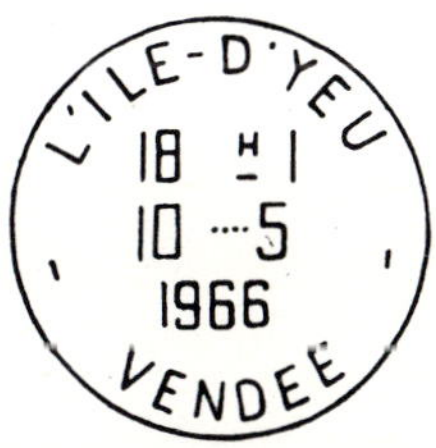

H14

CH1

O1

C2

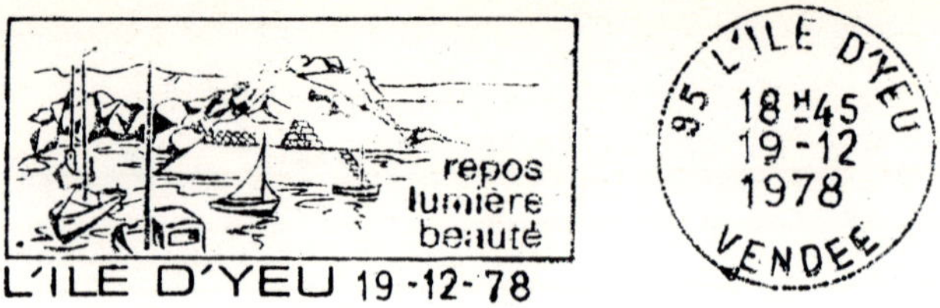
repos
lumière
beauté
L'ILE D'YEU 19-12-78
L'ILE D'YEU
18H45
19-12
1978
VENDEE

M3

ILE DE RE (Charente-Maritime)

The Ile de Ré is a favourite tourist centre boasting several attractive resorts and wooded settings. It is served from La Pallice and La Rochelle and today has a population of about nine thousand.

A letter dated 2 January 1626 is known from the Camp de L'ile de Ré. A post office was opened during the seventeenth century. A manuscript mark 'de Ré' was used on paid letters from 1727–29 and another 'de St. Martin de Ré' from 1730–32. Several handstamps are known from the second half of the eighteenth century including two oval types, both of which are rare and valuable. A letter of 1779 has a manuscript mark 'Deb. de L'Isle de Rhé'.

A forwarding agent cachet, known to us from 1819, is listed below and Kenneth Rowe's book on the subject lists three others: P. J. Michault used an oval cachet measuring 36 x 19 in black in 1826, Rivaille and F. Fhouillaud marked letters in manuscript in 1832 and 1833 respectively.

During the nineteenth century straight line types were used and then lozenges and the standard datestamps. Our list is certainly incomplete and we would welcome further information.

The island's capital is at St. Martin but offices were soon opened at other settlements: La Flotte (1752), Ars (1779), La Couarde (1852), St. Marie (1870s), Le Bois Plage (1888), Les Portes (1961), Rivedoux Plage (1971), Loix (1973).

Early Handstamps

EH1	1751 -64	**De Re**	Straight line	Black	150.00
EH2	1763 -80	**ISLE/DE/RHE**	Three lines in fancy oval	Black	150.00
EH3	1764 -92	**COLONIES/PAR/L'ISLE DE/RHE**	Four lines in fancy oval	Black	250.00

St. Martin : Handstamps

H1	1780 -91	**ST MARTIN EN RHÉ**	Straight line	Black	75.00
H2	1799	**16/ST. MARTIN EN RHÉ**	Two line	Black	50.00
H3	1802	**16/ST.MARTIN.DE.RÉ**	Two lines, 48mm	Black	35.00
H4	1819	As above	As above but 46mm	Black	25.00
H5	1829	As above	As above but 40mm	Black	25.00
H6	1818	**P.16P./ST MARTIN.DE.RÉ**	Two line Paid letter mark	Black	40.00
H7	1819 -37	**COLONIES PAR/ST.MARTIN-ISLE-DE-RÉ**	Two line	Black	100.00
H8	1837	**SAINT MARTIN-DE-RÉ/(16)**	Type 13	Black	25.00
H9	1846	**OUTRE-MER/St.MARTIN-DE-RE**	Double circle	Red	70.00
H10	1852	**3196**	Lozenge with small figures used with H11	Black	10.00 (H10–H11)
H11	1852	**ST MARTIN DE RÉ/16**	Type 15; used with H10	Black	

H12	1863	**3754**	Lozenge with large figures used with H13	Black	10.00
H13	1863	**ST MARTIN DE RÉ/16**	Single circle Type 16; used with H12	Black	
H14	1880s	**ST MARTIN DE RÉ/ CHARENTE INFERIERE**	Type 18	Black	8.00
H15	1900s	**ST MARTIN DE RÉ/ CHARENTE INFERIERE**	Type A	Black	5.00
H16	1979	**17 ST MARTIN. DE RÉ/ CHARENTE MARITIME**	Single circle	Black	1.00

: Machines

M1	1959	**ST MARTIN DE RÉ/CHARENTE MARITIME**	Single circle datestamp with five wavy lines to right	Black	1.50
M2	1960		Pictorial slogan. as above	Black	1.00
M3	1967	**17 ST MARTIN DE RÉ/ CHARENTE MARITIME**	Single circle datestamp with five wavy lines to right	Black	1.00
M4	1969	Datestamp: **17 ST MARTIN DE RÉ/CHARENTE MARITIME** Slogan: **St MARTIN/de Ré/ Page/d'Histoire/SES REMPARTS SON PORT**	Pictorial slogan to left of datestamp	Black	1.00
M5	1979	As M4	Slogan to right of datestamp	Black	0.75

: Official Marks

O1	1979	**CORRESPONDANCE/PHILATELIQUE**	Unframed	Black	1.00

La Flotte : Handstamps

H1	1752	**DELAFLOTE**	Straight line 20 mm. (A dry strike of 1770 has been recorded)	Black	150.00
H2	1786	**ILE DE RHÉ/LA FLOTTE**	Two line	Black	100.00
H3	1786 -91	**COLONIES/PAR/LA FLOTTE**	As EH3	Black	200.00
H4	1823	**COLONIES PAR/LA FLOTTE**	Two line	Black	150.00
H5	1829 -33	**PAYS D'OUTREMER/PAR LA FLOTTE**	Boxed	Black	250.00
H6	1841	**OUTRE-MER/LA FLOTTE**	As St Martin H9	Black	75.00
	1845			Red	90.00
H7	1795	**16/LA FLOTTE**	Two line 31mm long	Black	50.00

H8	1811	**As above**	As above 37mm	Black	40.00
H9	1820	**As above**	As above 38mm	Black	40.00
H10	1852	**1296**	Lozenge with small figures used with H11	Black	10.00
H11	1852		Type 15; used with H10	Black	
H12	1863	**1530**	Lozenge with large figures used with H13	Black	10.00
H13	1863		Type 16; used with H12	Black	
H14	1880s		Type 18	Black	10.00
H15	1900		Type A	Black	8.00

: Machines

M1	1960		Single circle datestamp without Department number with five wavy lines to right	Black	1.50
M2	1967	Slogan: **LA FLOTTE-ILE DE RÉ/Peche. Sports Nautiques.** Datestamp: **17 LA FLOTTE/CHARENTE-MARITIME**	Single circle datestamp. stamp with slogan to right	Black	0.75

Ars : Handstamps

H1	1786	**ARS. J. DE. RE**	Straight line	Black	150.00
H2	1818	**16/Ars/ST.MARTIN-DE-RE**	Three line, cursive	Black	100.00
H3	1852	**132**	Lozenge with small figure used with H4	Black	20.00
H4	1852		Type 15; used with H3	Black	
H5	1863	**176**	Lozenge with large figures used with H6	Black	20.00
H6	1863		Type 16; used with H5	Black	
H7	1880s		Type 18	Black	10.00
H8	1900s		Type A	Black	8.00
H9	1979	**17.ARS.EN.RE/CHARENTE MARITIME**	Single circle	Black	1.00

: Travelling Postmark

TP1	1905	**ARS-EN-RE A SABLANCEAUX**	Single circle with indented sides	Black	10.00

: Machines

M1	1962	**ARS EN RE/CHARENTE MARITIME**	Five undulating bars to right of datestamp	Black	1.50
M2	1967	**17 ARS EN RE/CHARENTE MARITIME**	Five undulating bars to right of datestamp	Black	1.00

La Couarde : Handstamps

H1	1852	**972**	Lozenge with small figures; used with H2	Black	40.00 (H1–H2)
H2	1852		Type 15	Black	
H3	1863	**1152**	Lozenge with large figures; used with H4	Black	30.00 (H3–H4)
H4	1863		Type 16	Black	
H5	1979	**17.LA COUARDE SUR MER/ CHARENTE.Mme**	Single circle	Black	1.00

: Machines

M1	1952	**LA COUARDE SUR MER/ CHARENTE MME**	Single circle datestamp with five wavy bars to right	Black	2.00
M2	1963	PTT Slogan	Datestamp as above, slogan to right	Black	3.00
M3	1967	**17 LA COUARDE SUR MER/CHARENTE Mme**	Single circle date-stamp with five wavy bars to right	Black	1.00
M4	1979	Datestamp: **17 LA COUARDE SUR MER/CHARENTE.Mme** Slogan: **LA COUARDE/ses Plages, ses Loisirs/LA COUARDE SUR MER date**	Datestamp to right of pictorial slogan	Black	0.75

: Official Marks

O1	1979	**CORRESPONDANCE/ PHILATELIQUE**	Unframed	Red	1.00

St Marie : Handstamps

H1	1870s		Type 17	Black	35.00
H2	1880s		Type 17	Black	35.00
H3	1900s		Type A	Black	25.00
H4	1979	**17 STE MARIE DE RE/CHARENTE. MARITIME**	Single circle	Black	1.00

: Machines

M1	1962	**STE MARIE DE RE/CHARENTE MARITIME**	Single circle date-stamp with five wavy bars to right	Black	1.50
M2	1967	**17 STE MARIE DE RE/CHARENTE MARITIME**	As above but date-stamp with department number	Black	1.00

Le Bois Place : Handstamps

H1	1979	**17.LE BOIS PLAGE EN RE/ CHARENTE Mme**	Single circle	Red	2.00

: Machines

M1	1953	**LE BOIS PLAGE EN RE/ CHARENTE MME**	Single circle date-stamp with five wavy lines to right	Black	1.50
M2	1953	Slogan: **VOIE-LUMIERE/SANTE-AIR IODE/MAI A SEPTEMBRE** Datestamp as above	Slogan to right	Black	2.50
M3	1967	**17.LE BOIS PLAGE EN RE/ CHARENTE MME**	Single circle date-stamp with five wavy lines to right	Black	0.75

Les Portes : Handstamp

H1		**17.LES PORTES/CHARENTE MARITIME**	Single circle	Black	1.00

: Machines

M1	1979	**17.LES PORTES EN RE/ CHARENTE MME**	Single circle datestamp with five undulating bars to right	Black	0.75

Rivedoux – Plage : Handstamps

H1	1979	**17.RIVEDOUX.PLAGE/ CHARENTE MME**	Single circle	Black	1.00

: Machines

M1	1979	**17 RIVEDOUX-PLAGE/ CHARENTE MME**	Single circle datestamp with five undulating bars to right	Black	0.75

: Official Marks

O1	1979	**CORRESPONDANCE/PHILATELIQUE**	Unframed	Red	1.00

Loix : Handstamps

H1	1979	**17 -LOIX/CHARENTE- MARITIME**	Single circle	Black	1.00

: Machines

M1	1979	**17 LOIX/CHARENTE MARITIME**	Single circle date-stamp with five bars to right	Black	0.75

Ile de Ré Cachets

C1	1819	**ST MARTIN/F.D. fils & B/ISLE DE RE**	Oval forwarding agent cachet	Black	20.00

C2	1826		Oval forwarding agent cachet used by P. J. Michault	Black	20.00
C3	1905	**HOTEL-RESTAURANT DES BALEINES ILE-DE-RE/ Vve DRUGEON**	Double oval	Violet	3.50
C4	1908	**HOTEL DES BALEINES/PORCHE/ HOTEL/du/Lion d'or/ILE DE RÉ**	Double circle	Red	3.50
C5	1914 –16	**PRISONNIERS DE GUERRE**	Straight line	Red or Violet	15.00
C6	1915	**VISA DE INTERPRÈTES**	Double circle with plant in centre	Red	20.00
C7	1917 –18	**VISA DES INTERPRÈTES/E.W.**	Double circle with letters in centre	Violet-black	18.00
C8	1917	**VISA DES INTERPRÈTES/F.C.**	As C7	Black	10.00
C9	1917	**VISA DES INTERPRÈTES/L.M.**	As C7	Violet or black	20.00
C10	1917	**VISA DES INTERPRÈTES/R.T.**	As C7	Violet	20.00
C11	1917	**Vu:/L'Interprète/M.S.**	Double oval	Violet	25.00
C12	1917	**VISA DE L'INTERPRÈTE/M.S.**	Double circle with letters in centre	Violet or blue-green	25.00
C13	1917 –18	**DEPOT DES PRISONNIERS DE GUERRE/ST.MARTIN DE RÉ/ VISA de L'interpréte**	Double circle	Red or blue-green	25.00
C14		**COLONIE TOIRAS/St MARTIN-de-RÉ/(Charente-Infre)**	Boxed	Black	3.00

1729 Entire from the Ile de Ré with manuscript marking 'de Ré'

Entire with EH2 'ISLE DE RHÉ' in oval with Fleurs de Lys

EH1

EH2

EH3

St Martin H1

St Martin H2

St Martin H3

St Martin H4

St Martin H5

St Martin H6

St Martin H7

St Martin H8

St Martin H9

St Martin H11

St Martin H13

La Flotte H2

La Flotte H4

La Flotte H7

La Flotte H8

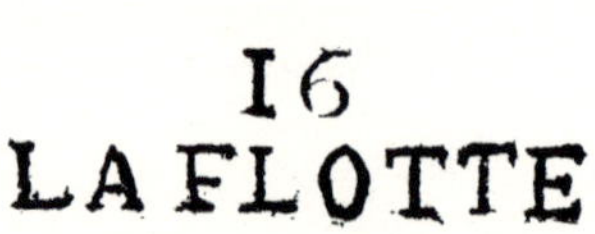

La Flotte H9

La Flotte M2

16
Ars
St MARTIN·DE·RE

Ars H2

Ars H9

Ars Travelling

ARS.J.DE·RE

Ars H1

La Couarde H5

les P.T.T vous offrent ....
DES
SITUATIONS D'AVENIR
renseignez-vous
dans les bureaux de poste

CORRESPONDANCE
PHILATELIQUE

La Couarde O1

Les Portes H1

Rivedoux Plage H1

Loix H1

C1

C3

C4

C5

C6

C11

C12

C13

C14

ILE D'AIX (Charente Maritime)

The Island of Aix (the 'x' is not sounded) lies about 4km off the Port of Fouras from which it can be reached by boat in about 25 minutes. It is also served by vessels from Rochefort, La Rochelle, Ile de Ré and Ile d'Oléron. The island is 3km long, has an area of 600m amd a present population of just over two hundred. Napoleon stayed on Aix from July 7–15 1815 and his house is one of the show places of the island. There is also a large fort built by Vauban.

Handstamps

H1	1813	**16/L'Isle-d'Aix/ROCHEFORT**	Three line	Black	80.00
H2	1830	**16/L'Isle d'Aix**	Two line	Black	60.00
H3	1832	**16/L'Isle d'Aix**	Two line	Black	60.00
H4	1839	**16/ILE D'AIX**	Two line	Black	45.00
H5	1849	**1545**	Lozenge with small figures	Black	40.00
H6	1863	**1833**	Lozenge with large figures	Black	35.00
H7	1849	**L'ISLE D'AIX/(16)**	Type 15	Black	40.00
H8	1870s	**L'ILE D'AIX/(16)**	Type 16	Black	18.00
H9	1880s	**L'ILE D'AIX/ CHARENTE INFERIERE**	Type 18	Black	10.00
H10	1900s		Type A	Black	15.00
H11	1929	**ILE-D'AIX/CHARENTE-INFR**	Single circle with broken ring	Black	3.00
H12	1951	**ILE D'AIX/CHARENTE MARITIME**	Single circle with broken ring	Black	2.00
H13	1979	**17 ILE D'AIX/CHARENTE MARITIME**	Single circle	Black	1.00

Machines

M1	1955	**ILE D'AIX/CHARENTE MARITIME**	Broken single circle date-stamp with five wavy bars to right	Black	2.00
M2	1955 –61	Slogan: **ILE D'AIX/ILE DE PAIX & DE REVE/HAUT LIEU NAPOLEONIEN/ SES MUSEES.** Datestamp as above	Broken single circle date-stamp with five wavy bars to right	Black	2.50
M3	1975	**17 ILE D'AIX/CHARENTE MARITIME**	Single circle datestamp with five wavy bars to right	Black	1.00

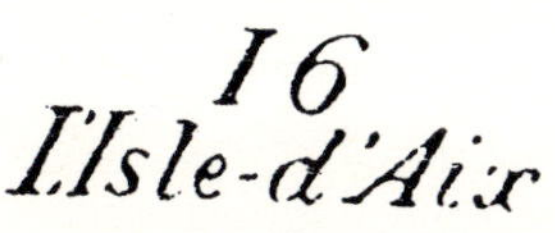

H2

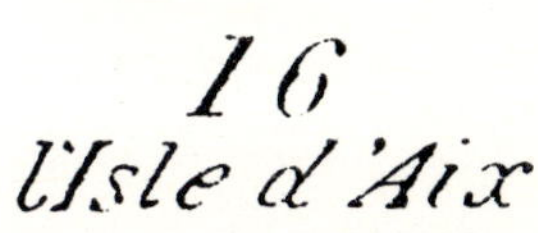

H3

H6

H7

H11

H12

H13

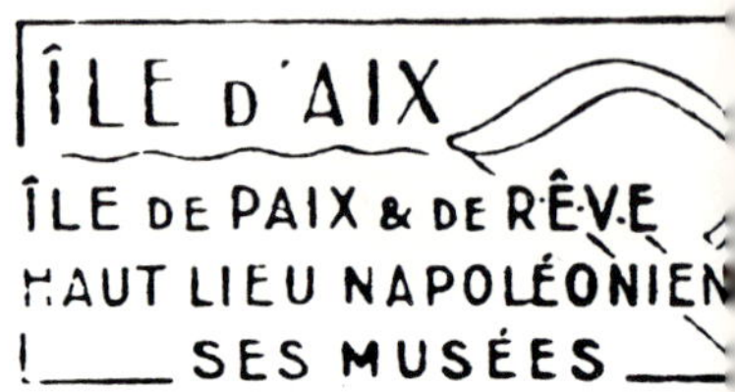

M2

ILE D'OLERON (Charente-Maritime)

Oleron in the second largest of the islands of France, exceeded only by Corsica. It is a popular holiday centre reached for many years by steamer from Rochfort. Since 1966 the island has been connected to mainland France by a bridge. The population is 14,000.

A post office was opened in 1767 and attractive three-lined marks were used up to about 1800, struck in red or black. These are scarce and valuable. From the 1850s when the lozenge came into use the office was known as Le Chateau d'Oleron. Other offices were opened at St Pierre (1838), St Georges (1848), St Denis (1869), St Trojan-les-Bains (1885), Dolus (1910), Boyardville (1915), La Cotiniere (1970), La Bree les Bains (1972).

Early Handstamps

EH1	1768 -90	**ISLE/DOLLE/RON**	Hexagon	Black	100.00
EH2	1770	**ISLE/DOLLE/RON**	Hexagonal	Red	400.00
			This mark became very mishapen over its period of use	Black	100.00
EH3	1813	**16/L'ISLE D'OLERON**	Two line	Black	40.00
EH4	1835	**ESPAGNE/PAR OLERON**	Two line	Black	60.00
EH5	1837	**ESPAGNE/PAR OLERON**	As above but larger letters	Red	50.00
EH6	1811	**P.16.P/L'ISLE D'OLERON**	Two line paid letter mark	Black	75.00

Le Chateau D'Oleron : Handstamps

H1	1852	**766**	Lozenge with small figures; used with H2	Black	10.00
H2	1852	**LE CHATEAU D'OLERON**	Type 15; used with H1	Black	
H3	1963	**916**	Lozenge with large figures used with H4	Black	10.00
H4	1863	**LE CHATEAU D'OLERON**	Type 16 used with H3	Black	
H5	1880s	**LE CHATEAU D'OLERON**	Type 17	Black	13.00
H6	1890s	**LE CHATEAU D'OLERON**	Type 18	Black	8.00
H7	1900	**LE CHATEAU D'OLERON**	Type A	Black	6.00
H8	c1900	**ENTREPOT DU CHATEAU D'OLERON**	Type A	Black	40.00

: Machines

M1	1955		Single circle datestamp with department number with five wavy bars to right	Black	1.50
M2	1968	**PTT** Slogan	Datestamp as above, Slogan as La Couarde, Ile de Ré, M2	Black	1.50
M3	1968		Single circle datestamp with dept number and five wavy bars to right	Black	0.75

St Pierre : Handstamps

H1	1828	**16/St Pierre d'Oleron/L'ISLE-D'OLERON**	Three line	Black	50.00
H2	1852	**3841**	Lozenge with small figures used with H3	Black	25.00
H3	1863		Type 15; used with H2	Black	
H4	1863	**3811**	Lozenge with large figures used with H5	Black	20.00
H5	1863		Type 16; used with H4	Black	
H6	1880s		Type 17	Black	18.00
H7	1890s		Type 18	Black	15.00
H8	1910	**ST PIERRE D'OLERON/ CHARENTE INFRE**	Single circle	Black	8.00
H9	1916	**ST PIERRE D'OLERON/ CHARENTE INFRE**	Type A	Black	8.00
H10	1979	**17 ST PIERRE D'OLERON PPAL/CHARENTE MME**	Single circle	Black	1.00

: Machines

M1	1959		Single circle datestamp without Department number with five wavy bars to right	Black	2.00

M2	1961		Pictorial slogan datestamp as above	Black	1.00
M3	1967		Single circle datestamp with Department number and five wavy bars to right	Black	1.00
M4	1976	Slogan: **ST PIERRE D'OLERON/ 'Au Centre d'L'ile'/PECHE/ TOURISME/COMMERCE** Datestamp: **17 ST PIERRE D'OLERON ppal/CH.MME**	Pictorial slogan to right of datestamp	Black	0.50

St Georges : Handstamps

H1	1852	**3054**	Lozenge with small figures; used with H2	Black	15.00
H2	1852		Type 15; used with H1	Black	
H3	1863	**3624**	Lozenge with large figures; used with H4	Black	10.00
H4	1863		Type 16; used with H3	Black	
H5	1880s		Type 17	Black	18.00
H6	1890s		Type 18	Black	15.00
H7	c1900		Type A	Black	8.00

: Machines

M1	1958	**ST GEORGES D'OLERON**	Single circle datestamp with five wavy bars to right	Black	1.50
M2	1959	Slogan: **SAINT-/GEORGES D'OLERON/SON-EGLESE, DES XIe ET XIIe/SES VINS/ SAINT GEORGES D'OLERON/ SES PLAGES/SES PINS**	Datestamp as above	Black	3.00
M3	1961	Slogan: **St GEORGES D'OLERON/ SES PLAGES-SES PINS/SON EGLISE DES XIe ET XIIe**	Datestamp as above	Black	1.50
M4	1964	**PTT Slogan**	Slogan as La Couarde, Ile de Ré M2, datestamp as above	Black	2.00
M5	1967	Datestamp: **17 ST GEORGES D'OLERON**	Slogan as M3 above	Black	2.00
M6	1967	**PTT Slogan**	Slogan as La Courde, ile de Ré M2, datestamp as M5 above	Black	2.50
M7	1979		M5 datestamp with five undulating bars to right	Black	0.75

St Denis : Handstamps

H1	1869		Type 17	Black	22.00
H2	1870s		Type 18	Black	18.00
H3	c1900		Type A	Black	15.00

: Machines

M1	1962	**St DENIS D'OLERON/ CHARENTE MME**	Single circle datestamp with five wavy bars to right	Black	1.50
M2	1967	**17 ST DENIS D'OLERON/ CHARENTE-MME**	Single circle datestamp with five wavy lines to right	Black	1.00

: Travelling Postmark

TP1	1910	**ST. DENIS D'OLERON A ST. TROJAN-LES-BAINS**	Single circle with indented sides	Black	17.00

St Trojan-Les-Bains : Handstamps

H1	c1900		Type A	Black	18.00
H2	1979	**17 ST TROJAN LES BAINS/CHARENTE MME**	Single circle	Black	1.00

: Machines

M1	1955	**ST TROJAN LES BAINS/ CHARENTE MME**	Single circle datestamp with five wavy bars to right	Black	2.50
M2	1956		Datestamp as above but with pictorial slogan	Black	1.50
M3	1964	Slogan: **ST TROJAN LES BAINS, Thalass-otherapie, Centre Oleron**	Datestamp as above with slogan to right	Black	1.00
M4	1967	**As above but datestamp 17 ST TROJAN LES BAINS/CHARENTE MME**	Slogan to right of datestamp	Black	0.75

Dolus : Handstamps

H1	1979	**17 DOLUS D'OLERON/ CHARENTE MARITIME**	Single circle	Black	1.00

: Machines

M1	1966	**DOLUS D'OLERON/ CHARENTE MME**	Five undulating bars to right of datestamp	Black	2.00
M2	1967	Datestamp: **17 DOLUS D'OLERON/CHARENTE MME**	Five undulating bars to right of datestamp	Black	0.75

Boyardville : Handstamp

H1	1979	**17 BOYARDVILLE/ CHARENTE MME**	Single circle	Black	1.00

La Cotiniere : Handstamps

H1	1979	**17 ST PIERRE. D'OLERON. LA COTINIERE/CHARENTE MME**	Single circle	Black	1.00

: Machines

M1	1965		Single circle date-stamp without Department number with five undulat-ing bars to right	Black	1.50

M2	1967	**17 ST.P.D'OLERON-LA COTINIERE/CH.MME**	Single circle date-stamp with five undulating bars to right	Black	0.75

La Brée Les Bains : Handstamps

H1	1979	**17 LA BREE.LES.BAINS/ CHARENTE MARITIME**	Single circle	Red	2.00

: Official Marks

O1	1979	CORRESPONDANCE/ PHILATELIQUE	Unframed	Black	1.00

Ile D'Oleron Cachets

C1	1862	**ADMINISTRATION DES LIGNES TELEGRAPHIQUES/STATION/ DE ST PIERRE/D'OLERON**	Double circle	Red	20.00
C2	1907	**PHARE de CHASSIRON/A.H.**	Single circle	Violet	8.00
C3	1917	**Depot des Prisonniers/DU CHATEAU D'OLERON/ Correspondance VERIFIÉE**	Boxed	Violet	25.00

The opening of the Pont D'Oléron was marked by the issue of a stamp and a slogan. Designed and engraved in Recess by C. Durrans the 25c stamp was in brown, greenish blue and blue (SG 1724, Yvert 1489); values: mint 0.10, used 0.08, FDC 50p. A slogan was used at nearby Bourcefranc.

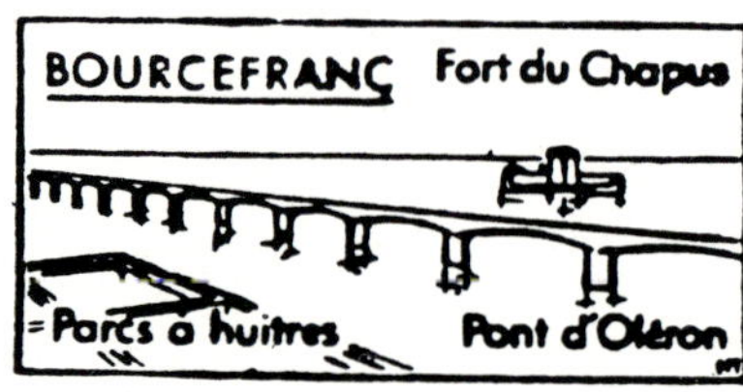

EH1

EH2

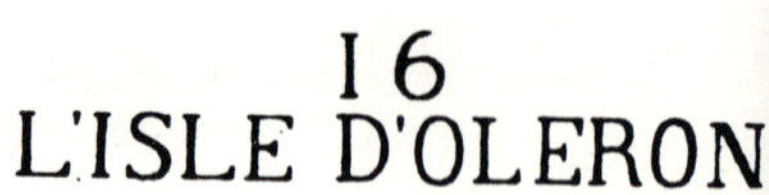

EH3

EH4

EH5

EH6

St Pierre H9

St Pierre H10

St Pierre M1

Dolus H1

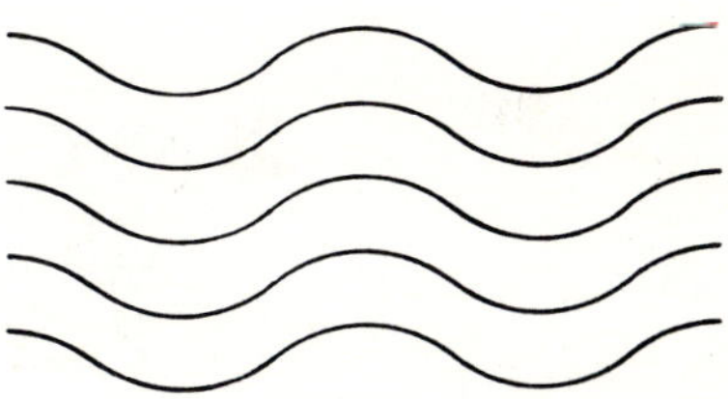

Dolus M2

La Cotinière H1

La Brée les Bairs H1

C3

C2

C1

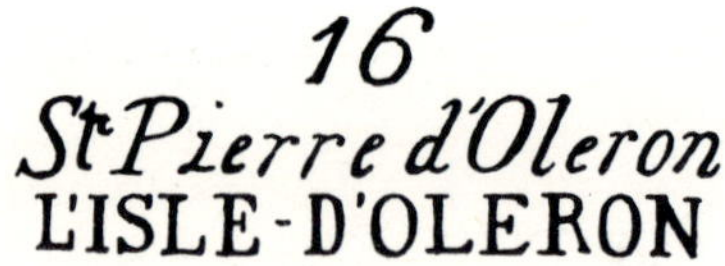

St Pierre H1

BIBLIOGRAPHY

A. Historical and Topographical Works

BARTHELEMY, GUY	*Iles.* A General guide to all the French Islands: Chausey, Bréhat, Sept-Iles, Batz, Ouessant, Molène Belle-Ile, Sein, Groix, Houat, Hoedic, Noirmoutier, Yeu, Ré, Aix, Madame and Oleron; also those off the Mediterranean coast: Porquerolles, Port-Cros, Levant and Lerins. Publications du Pelican, Clecy, 1976.
BARTHELEMY, GUY	*Les Iles Chausey,* Publications du Pelican, Clecy, 1973
BRUNET, ROGER	*Les iles, beautes de la France.* A full colour work on all the French Islands. Librairie Larousse, Paris, 1976.
CHAGNOLLEAU, JEAN	*Les Iles De L'Armor.* A general work on the islands of Brittany with many illustrations in colour. Editions des Horizons de France, Paris, 1911.
CHAUSEY	*Guide Touristique de l'Archipel des Iles Chausey.*
CALLOT	*Siege de L'Ile de Ré.* Dessins.
CASTAING, G.	*L'Ile de Ré.*
COMMERE, J.	*Les Iles Chausey,* 1965.
DESGRAVES, L.	*L'Ile d'Oleron,* Editions Melusine, La Rochelle, 1965.
DESROSEAUX, M.	*A L'Ile de Sein.*
DALIGAUT, M.	*Belle-Ile en Mer.*
FLEURIOT DE LANGLE	*L'Ile d'Aix,* 1949.
CAUSSENS, JEAN-GABRIEL	*L'Ile d'Yeu,* Editions Melusine, La Rochelle, 1969.
HERVE, ALAIN	*Belle-Ile,* 1950.
KEMMERER, Dr	*Histoire de L'Ile de Ré.*
JAMES, R. & SUIRE, L.	*L'Ile de Ré*
JOURDAN, Abbé	*Les Iles Chausey.*
LAOUENAN, ROGER	*Ile de Bréhat.*
LE BRAZ, A.	*Iles Bretonnes.*
LANCO, Y.	*Histoire de Belle-Ile,* 1954.
LEANDRE, LE GALLEN	*Belle-Ile,* 1906.
LE CUNFF, L.	*Sein,* 1958.
LE CUNFF, L.	*Les Iles de Glénan.*

LE RAY, LOUIS — *Belle-Ile en Mer.*
LUCAS, ALBERT — *Ouessant, Ile sur Ponant.*
MARTIN, HENRI — *Au Pays Noirmoutrin.*
MARTINEAU, G. — *L'Ile de Ré.*
MINOT — *L'Ile aux Moines.*
PAGNIEZ, Y. — *Ouessant, 1965.*
PARIS — *Histoire de Belle-Ile en Mer,* 1870.
PIET, F. — *Recherches sur L'Ile de Noirmoutier.*
PRASTEAU, JEAN — *Les Iles d'Ouest,* 1954.
RATCLIFFE, M. — *Cézembre, Histoire de l'Ile.*
QUATREFAGES, A. DE — *Sejour à Chausey,* 1842.
RAIMOND, J. — *L'Ile de Noirmoutier. Editions Melusine, La Rochelle 1967.*
RICHARD, O. — *L'Ile d'Yeu autrefois.*
RICHARD, S. — *Sein, Ile des Trepasses.*
RICHARD, R. P. — *Ile de Sein.*
THOMAS, P. — *L'Ile d'Oleron à travers les siecles.*
THOMAS — *L'Ile d'Ouessant,* 1948.
TOUCHE, CHARLES de — *Histoire de Belle-Ile.*
VIAU-GRAND-MARAIS — *Guide de Noirmoutier.*
VIAU-GRAND-MARAIS — *Guide d'Yeu,* 1946.
VIAU-GRAND-MARAIS — *Causeries Noirmoutrines.*
VALET — *L'Ile de Bréhat,* 1948.
VINCENT, Dr — *L'Ile de Groix autrefois,* 1895.
YBER, LUC — *L'Ile Bréhat,* 1953.

Selection of Works on Mont St. Michel

ADAMS, HENRY — *Mont Saint-Michel & Chatres,* 1905. Last edition 1959.
CHAUVEAU, G. M. — *Mont St. Michel.* In various languages including English.
BASTIEN, A. P. — *Le Mont Saint-Michel au Peril de la Mer.*
CARTON, JACQUES — *The Sights of Mont Saint-Michel.*
GOUT, PAUL — *Le Mont-Saint-Michel.* The standard work on the history, archeology and architecture of the Abbey and the Town, in four volumes (Tome 1, parts 1 and 2, Tome 2, parts 1 and 2). Librarie Armand Colin, Paris 1910.
MAUCLAIR, C. — *Le Mont Saint-Michel.* 1927.

MAUCLAIR, C. — *Mont Saint-Michel* (Translated by John Gilmer).

GOUT, PAUL — *Tourist's Guide to Mont St. Michel,* 1919.

CORROYER, E. — *Descriptive Guide of Mont Saint-Michel,* 1883.

TOMBELAINE, Marquis of — *St. Michael's Mount and its Marvels.*

B. Philatelic Works

BARGAIN, L. — *La Poste à L'Ile-Tudy, une longue histoire,* article in 'Bulletin Municipal', 1978.

BRUNO — *Catalogue de la Vendée.*

CARNEVALE-MAUZAN, M. — *Les camps de prisonniers de guerre et internés civile en France et en Afrique du Nord 1914–19; les marques de censure,* Grenoble 1967.

DELOSTE, C. — *Histoire Postal et Militaire de la Guerre de 1914–18* Bischwiller, 1968.

DANAN, Y. M. — *Emissions locales et affranchissements de Guerre des Iles de la Manche,* Pub. Le Monde des Philatelistes, 1968.

Histoire Postale des Iles de la Manche, Tome 1, *Les Affranchissements de Guerre de 1870 à 1944,* Pub. Le Monde des Philatelistes, Etude 195, 1975.

DUBUS & MICOULSKY. — *Marques Postales 1969–1876, La Manche (48),* Amiens, 1945.

DUBUS, LA — *Marques Postales 1698–1876, La Charente-Inferieure,* Royan, 1958.

FORRESTER, R. — *The French Connection,* two articles in 'Bulletin of the Channel Islands Specialists Society', No. 148, 1975, p. 9–10, No. 150 1976, p. 7–8.

LAVARACK, D. — *Nomenclature des Bureaux de Poste Français,* 1876–99; Boulogne, 1967.

LENAIN, L. — *La Poste de l'Ancienne France des origines à 1791.*

NEWPORT, O. W. — *Postal History and Markings of Mont Saint-Michel,* 'Stamp Collecting', 11 Feb. 1966, p. 863–4.

Stamps and Postal History of the Channel Islands, Heinemann, 1972.

Specialised Priced Catalogue of Channel Islands Stamps, Channel Islands Specialists' Society, Tenth Edition, June 1976.

Priced Cataloque of Channel Islands Postal Markings, Channel Islands Specialists' Society, Fifth Edition, January 1976.

Editorial, Les Iles Normandes, Journal of the Channel Islands Specialists' Society, Vol. 1 No.4, 1977.

POTHION, J. *Catalogue des Cachets Courriers-Convoyeurs-Lignes 1877–1966,* Paris, 1972.

Catalogue des Grilles et Petits Chiffres des Bureaux de Province (1849–62), Paris 1974.

Nomenclature des Bureaux de Postes Français, 1852–76 Petits & Gros Chiffres Cotes, Paris 1877.

Dictionnaire des Bureaux de Poste Français, 1904–14, Paris 1977.

POTHION, J. *Dictionnaire des Bureaux de Poste Français,* 1575–1904, Paris, 1976.

WHITNEY, J. T. *The French Channel Islands.* Some preliminary notes on their Postal History and Markings, 1977.